W9-CKL-223

• ENERGY

• SIMULATION-TRAINING

• OCEAN ENGINEERING AND INSTRUMENTATION

Research Papers of the Link Foundation Fellows

Volume 3

Brian J. Thompson
Editor

50th Anniversary
1953-2003

Published by
The University of Rochester Press
in Association with
The Link Foundation

First published 2003

University of Rochester Press
668 Mt. Hope Avenue
Rochester, New York, 14620, USA

and at

PO Box 9, Woodbridge,
Suffolk IP12 3DF, UK

ISBN 1-58046-124-7

Library of Congress Cataloging-in-Publication Data

Energy, simulation-training, ocean engineering and instrumentation : research papers of the Link Foundations Fellows / Brian J. Thompson, editor.
p. cm.
ISBN 1-58046-124-7 (alk. paper); Volume 3
1. Power resources 2. Energy development. I. Thompson, Brian J., 1932–

TJ 163.2.E4867 2001
333.79—dc21 2001021331

This publication is printed on acid-free paper
Designed and Typeset by Straight Creek Bookmakers
Printed in the United States of America

The editor and the authors dedicate this volume to the memory of Edwin A. Link and his innovative and entrepreneurial spirit.

TABLE OF CONTENTS

PART III. OCEAN ENGINEERING AND INSTRUMENTATION

PREFACE

An important ingredient of education at the Ph.D. level is the reporting of the original scholarly and intellectual content of the research conducted by the candidate for the degree. Thus the Link Foundation asks that each recipient of a Link Foundation Fellowship prepare and submit a journal length article resulting from their work under the Link Foundation sponsorship. The Link Foundation believes in this discipline and has decided to complete this process by the formal publication of these papers in a single volume.

The current fellowship programs are in the fields of energy, simulation-training and ocean engineering and instrumentation. The Energy fellowships have been offered since 1983 and the papers of these fellows have been published in a series of fifteen volumes from 1986–2000 under the title "Research Reports of the Link Energy Fellows." Simulation-Training fellows have submitted research papers since the inception of the program in 1990; these papers have been published informally until 2000. Finally the program in Ocean Engineering and Instrumentation was launched in 1998. Starting with Volume 1 in 2001, the research papers of the Link Foundation Fellows are published together in three part volumes. Volume 3 is presented here and contains eleven papers submitted by the 2001–2002 Fellows.

The Link Foundation is pleased to have been able to support the students represented here. The work reported covers a wide variety of research topics carried out at leading universities and colleges.

Brian J. Thompson
University of Rochester and
The Link Foundation
Editor

PART I

ENERGY

Synthesis and Characterization of $La_{13}V_4Cu_9O_{38.5}$ and Its Decomposition Product $La_{14}V_6CuO_{36.5}$: An N-Type Transparent Conductor

Z. Serpil Gönen

Department of Chemistry and Biochemistry

University of Maryland

College Park, MD 20742

Research Advisor: Bryan W. Eichhorn

ABSTRACT

The search for new p-type transparent conductors to prepare transparent p-n junctions and to form new solar cells led to the synthesis of $La_{14}V_6CuO_{36.5}$. This compound is transparent and contains $(CuO_2)^{3-}$ linear sticks that are diluted in a lanthanum oxide matrix. The estimated band gap from the diffuse reflectance optical measurements conducted on a polycrystalline sample is ~2.9 eV. Seebeck effect measurements, done on the thin films, show that the carriers are electrons and the compound is an n-type conductor regardless of the annealing conditions. $La_{14}V_6CuO_{38.5}$ is originally obtained as a decomposition product of $La_{13}V_4Cu_9O_{36.5}$ at 1200°C. Therefore, the structure and physical properties of $La_{13}V_4Cu_9O_{36.5}$ and $La_{14}V_6CuO_{38.5}$ are presented.

INTRODUCTION

There is a current interest in developing renewable energy resources such as photovoltaic electric generators. A great stress has been placed on development of high efficiency, low cost solar cells to respond to the potential demand in the power generation market.

Doped versions of In_2O_3, SnO_2 and ZnO are often used in solar cells as transparent n-type conductors. These compounds are transparent due to their large band gaps (~3eV) but conductive since they are heavily n-doped. Despite the importance of n-type transparent conductors, there are no analogous technologically-viable p-type transparent conductors known. If new p-type transparent conductors could be developed, they could be used in preparing transparent p-n junctions and allow for new solar cells with n-type harvesting layers. Therefore, new materials are necessary for further applications development in this area.

The Cu(I) based delafossite, $CuAlO_2$, has been used as a prototype for the thermally stable, optically transparent p-type conductors but their conductivities are low [1]. The linear $Cu(I)O_2$ sticks are the crucial feature in the delafossites since they are responsible for the p-doping. Although the Cu^+ ion is quite common in molecular inorganic chemistry $Cu(I)O_2$ sticks are rare in solid state compounds [2]. Besides the $CuMO_2$ delafossites (M = Al, Fe, Sc, Co, Rh, Ga, Y or lanthanide) [3–7], this unit has been identified in several compounds including the parent binary oxide, Cu_2O [8–9], two high Tc superconductivity-related cuprates, $YBa_2Cu_3O_6$ [10] and $Pb_2Sr_2(Ca,Y)Cu_3O_8$ [11], $SrCu_2O_2$ [12] and $Sr_5(VO_4)_3(CuO)$ [13]. The ternary alkali metal copper oxides with the general formulae, ACuO (A = Li–Cs) [14], $A_3Cu_5O_4$ (A = K–Cs) [15] and A_3CuO_2 (A = Na–Rb) [16] represent a distinct class of Cu(I) oxide system where the $[OCuO]^{3-}$ unit is bent ($170° < O\text{-}Cu\text{-}O < 180°$). $Na_5[CuO_2](OH)_2$ also belongs to this class with (O-Cu-O) ~ 170° [17]. Interestingly, some of these compounds have potential or proven electronic applications [18–19]. For example, $SrCu_2O_2$ is another example of transparent p-type conductor, with low conductivity, that contains $Cu(I)O_2$ sticks [12].

The key feature in the $CuMO_2$ delafossites is the presence of the linear $[OCuO]^{3-}$ sticks that are diluted with MO_6 octahedra. The latter gives the compounds their transparent nature whereas the $[OCuO]^{3-}$ sticks provide p-type conductivity.

Here the synthesis, structure and optical spectrum of a new copper vanadate, $La_{14}V_6CuO_{36.5}$, that contains isolated VO_4^{3-} tetrahedra and $[OCuO]^{3-}$ sticks are described. It is an example of a Cu(I) oxide and transparent Cu-O material and shows n-type conductivity regardless annealing conditions. This compound was first obtained as a decomposition product of $La_{13}Cu_9V_4O_{38.5}$. The formation of hexagonal $La_3Cu_2VO_9$ has been reported by Jansson et al. [20]. However our investigations of this compound showed the presence of a superstructure that can be best described with the stoichiometry, $La_{13}Cu_9V_4O_{38.5}$. Therefore the structure and physical properties of $La_{13}Cu_9V_4O_{38.5}$ will also be presented.

EXPERIMENTAL

All the chemicals were of high purity (> 99 %) purchased from Cerac or Alfa Aesar. La_2O_3 was dried at 950°C before each use. Powder X-ray diffraction (XRD) patterns were recorded at 25°C using a Bruker D8 diffractometer (CuK_a radiation). XRD data were collected in 48 h runs using a step width of 0.02° between $10° < 2q < 90°$ for Rietveld refinement.

Chemical compositions of the nominal "$La_3Cu_2VO_9$" and $La_{14}V_6CuO_{36.5}$ samples were determined by wavelength dispersive spectroscopy (WDS) analysis using a JEOL JXA-8900 microprobe analyzer. $LaPO_4$, V and Cu were used as standards for the analysis of $La_{14}V_6CuO_{36.5}$ and $La_{14}V_6CuO_{36.5}$ was used for $La_3Cu_2VO_9$. Typically 10 measurements were performed for each compound and the average and the standard deviation was calculated using EXCEL. For each compound the total atom percentage for all non-oxygen elements were normalized to 100%. The average atom percentage for each element was multiplied by the oxidation state of the element and the sum of these numbers was divided by the sum of the oxidation states. Then, the average atom percent was divided by this number to obtain the number of moles of the element in the formula unit. The same calculations were done using the standard deviation rather than the average for each element to calculate the errors.

SYNTHESIS

$La_3Cu_2VO_9$ was prepared by reacting a stoichiometric mixture of the constituent oxides, La_2O_3, V_2O_5 and CuO (total mass: 2g) at 1000°C for 24 h and then at 1025°C for 48 h. WDS analysis of the resulting product showed that the La : V : Cu ratio was 13 : 4 : 9. Accordingly, a sample corresponding to the formula $La_{13}Cu_9V_4O_{38.5}$ was prepared by reacting stoichiometric amounts of the constituent oxides at 1000 °C for 24 h and then at 1025°C for 48 h. The black polycrystalline product, $La_{13}Cu_9V_4O_{38.5}$, was single phase and free from starting materials.

$La_{14}V_6CuO_{36.5}$ was first obtained as a high temperature (1100°C) decomposition product of $La_3Cu_2VO_9$. Later it was prepared from the stoichiometric mixture of La_2O_3, V_2O_5 and CuO in the molar ratio of 7 : 3 : 1. The mixture was loaded into an alumina crucible and heated for 36 h at 1100°C. After cooling to room temperature (5°C/min), the mixture was reground and refired at 1100°C for an additional 12 h. The sample was then cooled to room temperature at a rate of 1°C / min. The resulting product was a homogeneous pale green crystalline solid and it was single phase.

CHARACTERIZATION

Rietveld analysis of $La_{13}Cu_9V_4O_{38.5}$: Unit cell parameters were obtained by least-squares refinement of powder XRD data corrected for sample displacements. Rietveld profile analyses were carried out (Riqas, MDI) using split Pearson VII profile shape functions. Initial models for $La_{13}Cu_9V_4O_{38.5}$ were generated from the atomic coordinates for $La_4Cu_3MoO_{12}$ [21] and the refined cell constants of the compound. Structures were determined by sequentially refining the background coefficients, structure factors, lattice parameters and peak shape parameters. Isotropic thermal parameters and the positional parameters for the heavy atoms were refined in the last two cycles. Oxygen atomic positions were in positions associated with the heavy atoms and refined together for $La_{13}Cu_9V_4O_{38.5}$.

Crystallography for $La_{14}V_6CuO_{36.5}$: A colorless block with dimensions 0.163 x 0.132 x 0.116mm^3 was placed and optically centered on the Bruker SMART CCD system at –80°C. The initial unit cell was indexed using least-squares analysis of a random set of reflections collected from three series of 0.3° wide w scans (25 frames/series) that were well distributed in reciprocal space. Data frames were collected [MoKa] with 0.3° wide w-scans, 15 seconds/frame, 606 frames per series. Five complete series were collected providing a complete sphere of data to $2q_{max} = 55°$. A total of 13641 reflections were collected and corrected for Lorentz and polarization effects and absorption using Blessing's method as incorporated into the program SADABS with 4110 unique [R(int) = 0.0324].

The SHELXTL program package was implemented to solve and refine the structure. System symmetry and lack of systematic absences indicated the possible space groups to be either the centrosymmetric space group $P\overline{1}$ (no.2) or the acentric space group $P1$ (no. 1) with intensity statistics indicating the former. The structure was determined by direct methods with the successful location of nearly all atoms using the program XS. After locating all atoms and refining with anisotropic thermal parameters, it became evident that one of the vanadium atoms was disordered over two sites and was modeled as such. The resulting occupancies were V1a:V1b, 0.23:0.77. Oxygen O(20) was also found to be half occupied. The final structure was refined to convergence [D/s £ 0.001] with R(F) = 2.81 %, wR(F^2) = 6.25 %, GOF = 1.083 for all 4110 unique reflections [R(F) = 2.57 %, wR(F^2) = 6.13 % for those 3844 data with Fo > 4s(Fo)]. A final difference-Fourier map possessed many large peaks near the heavy atoms but these were considered ghosts due to the highly absorbing nature of the compound (m = $18.35 mm^{-1}$).

Magnetic measurements: Variable temperature d. c. magnetic susceptibilities were measured using a Quantum Design MPMS SQUID magnetometer. Polycrystalline samples, loaded into gelatin capsules, were cooled down to 5 K (ZFC, zero field cooled) and then a field of 5000 Oe was applied. The change in magnetization was measured as the samples were warmed from 5 K to 350 K. The field cooled (FC) measurements were carried out by cooling samples in a 5000 Oe field and then measuring the magnetization as the samples were warmed in the same field.

Resistivity Measurements: D. C. resistivities were measured as a function of temperature (20–300 K) using the standard four-probe method on rectangular blocks cut from sintered pellets.

Diffuse Reflectance: The spectrum was obtained at DuPont CR&D, DE.

RESULTS

$La_4Cu_3MoO_{12}$ crystallizes in a hexagonal $YAlO_3$–type structure where Cu^{2+} and Mo^{6+} cations have a trigonal bipyramidal oxygen coordination [21]. It is shown here that $La_3Cu_2VO_9$ is isostructural with $La_4Cu_3MoO_{12}$ and there is a superstructure that can be best described with the stoichiometry, $La_{13}Cu_9V_4O_{38.5}$. WDS analysis showed that the V:Cu ratio was 1.0 : 2.28. This ratio is close to the one (1 : 2) reported by Jansson et al. [20] on the basis of a EDS analysis. Normalizing the transition metal content to 13 (i. e., the number of transition metal atoms per layer of the unit cell), WDS data give a 4 : 9 ratio for V : Cu. Through comparison with the WDS data on a $La_{14}V_6CuO_{36.5}$ single crystal standard, the actual formula is calculated to be $La_{13.1(2)}Cu_{9.0(3)}V_{3.9(2)}O_{38.5}$ for the nominal $La_3Cu_2VO_9$ sample.

"$La_3Cu_2VO_9$" is black, homogeneous, essentially single-phase material. It phase-separates at 1200°C to give a crystalline Cu(I) phase and other binary (CuO) and ternary (La_2CuO_4) phases. Single crystal X-ray diffraction study showed that the colorless Cu(I) oxide was $La_{14}V_6CuO_{36.5}$.

Structures

$La_{13}Cu_9V_4O_{38.5}$: Inspection of the powder pattern reveals that the structure of "$La_3Cu_2VO_9$" is similar to that of hexagonal $La_4Cu_3MoO_{12}$ (a_h = 4.0045(6) and c_h = 10.6791(5) Å) reported by Poeppelmeier *et al.* [21]. However, indexing of the powder pattern of Cu/V phase showed the presence of $a = \sqrt{13}a_h$ superstructure where a » 14.4 Å. The latter is in excellent agreement with the electron diffraction data of Jansson *et al.* [20]. Based on these observations, different models were constructed in order to solve the

structure of $La_3Cu_2VO_9$. The best refinement was obtained with a model was generated using the supercell parameters reported [1] for $La_3Cu_2VO_9$ and the general structural features of hexagonal $La_4Cu_3MoO_{12}$.

The relation between the larger cell (Jansson et al.) [20] and the smaller hexagonal cell of Poeppelmeier *et al.* [21] is $a = \sqrt{13}a_h$ and $c = c_h$, where a_h and c_h are the lattice parameters of the smaller hexagonal cell (Figure 1). Based on the systematic absences in the XRD data and the symmetry of the proposed superstructure, the space group $P6_3/m$ was chosen for the construction of this model. The unit cell of the superstructure contains 26 metal atoms per layer (13 La and 13 transition metals) in which the transition metals reside on five independent crystallographic positions in the asymmetric unit (Wyckoff positions 6h´4 and 2d´1).

Several structural models for $La_{13}Cu_9V_4O_{38.5}$ were constructed wherein Cu and V were distributed on the four 6h sites and the single 2d site in the cell, while maintaining the generic Poeppelmeier structure. Partial ordering of Cu and V on some (or all) of the sites was necessary to obtain the $\sqrt{13}a_h$ superstructure. The best refinements were obtained with V on 2d sites and Cu and V ordered in various ways on the four 6h sites. The observed,

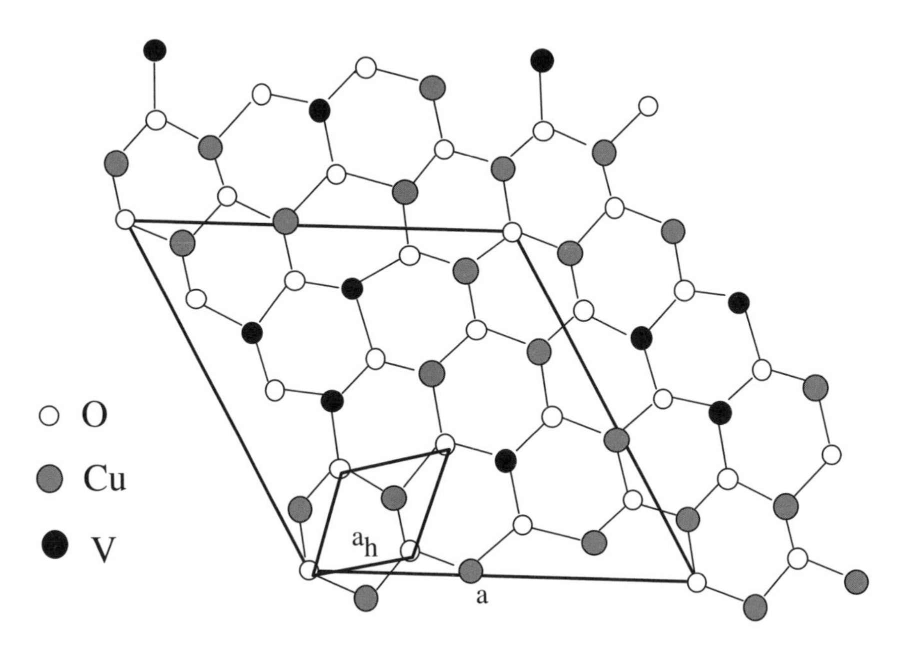

Figure 1. Proposed ordered hexagonal superstructure of $La_{13}Cu_9V_4O_{38.5}$ (large cell, a) showing the relationship to the smaller hexagonal $La_4Cu_3MoO_{12}$ cell (a_h). The projection is on the a-b plane.

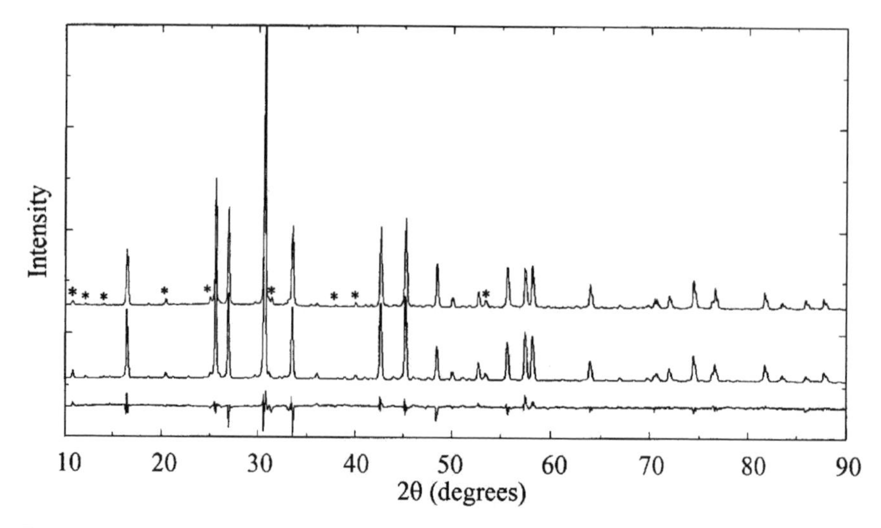

Figure 2. Observed (top), calculated (middle) and difference (bottom) profiles of powder XRD data for $La_{13}Cu_9V_4O_{38.5}$. *'s denote the peaks that can only be indexed in the $\sqrt{13}a_h$ hexagonal superstructure.

calculated and difference profiles of $La_{13}Cu_9V_4O_{38.5}$ are given in Figure 2. The $\sqrt{13}a_h$ superstructure reflections are marked with asterisks. The R factors and the refined unit cell parameters are given in Table 1.

During the course of this study, Vander Griend et al. published a paper describing the neutron and electron diffraction studies done on $La_3Cu_2VO_9$ [22,23]. The structure they described closely resembles the one shown in Figure 1. ED studies [22] agree with the $P6_3/m$ space group and show the presence of a $\sqrt{13}a_h$ superstructure. However, they reported the stoichiometry of the compound as $La_3Cu_2VO_9$ [23].

Table 1. Unit cell parameters and R factors for $La_{13}Cu_9V_4O_{38.5}$.

	$La_{13}Cu_9V_4O_{38.5}$
Space group	$P6_3/m$
a (Å)	14.4371(2)
b (Å)	14.4371(2)
c (Å)	10.6786(5)
V (Å^3)	1927.53(9)
d (g/cm^3)	5.5227(4)
R (%)	9.67
Rwp (%)	10.05

$La_{14}V_6CuO_{36.5}$: $La_{14}V_6CuO_{36.5}$ forms cleanly form stoichiometric ratios of the binary oxides at 1100 °C but also crystallizes as a decomposition product of $La_3VCu_2O_9$ at high temperatures. The individual crystallites are transparent, very pale green needles (essentially colorless) and were characterized by WDS and single crystal X-ray diffraction. The semi-quantitative WDS analysis confirmed the presence of La, V and Cu and gave compositions (atomic %; La, 25.7; V, 9.0; Cu, 1.8) in good agreement with the formula determined from the single crystal structure determination (atomic %; La, 24.3; V, 10.4; Cu, 1.7).

$La_{14}V_6CuO_{36.5}$ is triclinic, P-1, and contains seven La^{3+} ions, three V^{5+} ions and 0.5 Cu^+ ions in the asymmetric unit. A summary of the crystallographic data is given in Table 2 and listing of fractional coordinates and selected bond distances are given in Tables 3 and 4. A view of the unit cell where the CuO_2 units are highlighted is shown in Figure 3. The structure contains isolated VO_4^{3-} tetrahedra and linear $(OCuO)^{3-}$ units linked together by a lanthanum oxide network. The VO_4^{3-} tetrahedra are quite regular for V(2) and V(3). The V(1) atom is disordered over two sites that are separated by 0.817(5) Å. The percentage of occupancy for the two sides is 75:25%. O(19) is also partially occupied and is refined at 75%. The O(20) atoms are only 0.699 Å apart and are only 50% occupied. The remainder of the structure is crystallographically well behaved. Three of the lanthanum ions, La(2), La(3)

Table 2. Crystallographic data for $La_{14}V_6CuO_{36.5}$.

Empirical formula	$La_{14}V_6CuO_{37}$	
Temperature	173(2) K	
Wavelength	0.71073 Å	
Crystal system	Triclinic	
Space group	P-1	
Unit cell dimensions	a = 9.7375(11) Å	a= 88.083(2)°.
	b = 9.9167(11) Å	b= 64.141(2)°.
	c = 10.3684(12) Å	g = 87.847(2)°.
Volume	900.16(18) Å^3	
Z	1	
Density (calculated)	5.361 g/cm^3	
Absorption coefficient	18.350 mm^{-1}	
F(000)	1261	
Theta range for data collection	2.06 to 27.50°.	
Independent reflections	4110 [R(int) = 0.0324]	
Absorption correction	Empirical, Sadabs	
Refinement method	Full-matrix least-squares on F^2	
Goodness-of-fit on F^2	1.083	
Final R indices [I>2sigma(I)]	R1 = 0.0257, wR2 = 0.0613 [3844 Data]	
R indices (all data)	R1 = 0.0281, wR2 = 0.0625	

Table 3. Atomic coordinates (x 10^4) and equivalent isotropic displacement parameters (Å^{2}x 10^3) for $La_{14}V_6CuO_{36.5}$.

	x	y	z	U(eq)	occupancy
La(1)	6598(1)	6889(1)	554(1)	4(1)	1
La(2)	6454(1)	9960(1)	2740(1)	6(1)	1
La(3)	9009(1)	10001(1)	–1273(1)	4(1)	1
La(4)	8929(1)	3378(1)	987(1)	4(1)	1
La(5)	9208(1)	6753(1)	–3493(1)	5(1)	1
La(6)	8234(1)	6465(1)	3250(1)	11(1)	1
La(7)	3683(1)	6836(1)	4627(1)	15(1)	1
Cu(1)	5000	10000	0	13(1)	1
V(2)	7935(1)	3622(1)	–2164(1)	5(1)	1
V(3)	8027(1)	10024(1)	5135(1)	5(1)	1
V(1A)	5082(5)	6245(4)	–3275(5)	6(1)	0.75
V(1B)	5280(2)	6599(1)	–2663(2)	6(1)	0.25
O(1)	10649(5)	6018(4)	–6147(4)	9(1)	1
O(2)	3833(5)	9560(4)	4569(4)	10(1)	1
O(3)	8689(5)	9057(4)	3579(4)	9(1)	1
O(4)	8753(4)	5883(4)	941(4)	5(1)	1
O(6)	6487(5)	9388(4)	520(4)	6(1)	1
O(7)	6227(5)	7575(4)	2966(4)	7(1)	1
O(8)	3616(5)	7053(4)	2227(4)	11(1)	1
O(9)	7533(5)	5148(4)	–1383(4)	12(1)	1
O(10)	8176(5)	9274(4)	–3450(4)	11(1)	1
O(11)	11221(5)	9189(4)	–1019(4)	6(1)	1
O(12)	11474(5)	5904(4)	–3361(4)	7(1)	1
O(13)	8975(5)	7632(4)	–1373(4)	5(1)	1
O(14)	8814(5)	2603(4)	–1346(4)	11(1)	1
O(15)	8775(5)	11590(4)	4789(4)	11(1)	1
O(16)	4065(6)	7615(4)	–3243(5)	15(1)	1
O(17)	4203(6)	5356(6)	–1768(6)	34(1)	1
O(18)	7022(6)	6233(8)	–4042(6)	45(2)	1
O(19)	5705(10)	7407(7)	–1508(8)	53(2)	0.75
O(20)	5277(18)	5054(17)	–5352(12)	68(5)	0.5

and La(7), are seven coordinate monocapped trigonal prisms whereas the remaining ions are eight-coordinate bicapped trigonal prisms.

The copper atom in the linear $(OCuO)^{3-}$ ion resides on an inversion center and has a Cu-O distance of 1.831(4) Å, which is typical for two-coordinate Cu(I). The bond valence for the copper atom, +1.07, supports the oxidation state assignment. The $(OCuO)^{3-}$ ions are relatively isolated in the lanthanum oxide matrix and connect adjacent LaO layers.

Table 4. Selected bond lengths for $La_{14}V_6CuO_{36.5}$.

Cu(1)-O(6)	1.831(4)
Cu(1)-O(6)′	1.831(4)
V(2)-O(9)	1.692(4)
V(2)-O(8)	1.702(4)
V(2)-O(14)	1.727(4)
V(2)-O(1)	1.727(4)
V(3)-O(10)	1.683(4)
V(3)-O(15)	1.700(4)
V(3)-O(2)	1.735(4)
V(3)-O(3)	1.757(4)
V(1A)-V(1B)	0.831(5)
V(1A)-O(16)	1.644(6)
V(1A)-O(17)	1.658(7)
V(1A)-O(18)	1.700(7)
V(1A)-O(20)	2.10(2)
V(1A)-O(20)	2.42(2)
V(1B)-O(17)	1.628(5)
V(1B)-O(19)	1.664(6)
V(1B)-O(18)	1.711(6)
V(1B)-O(16)	1.808(5)

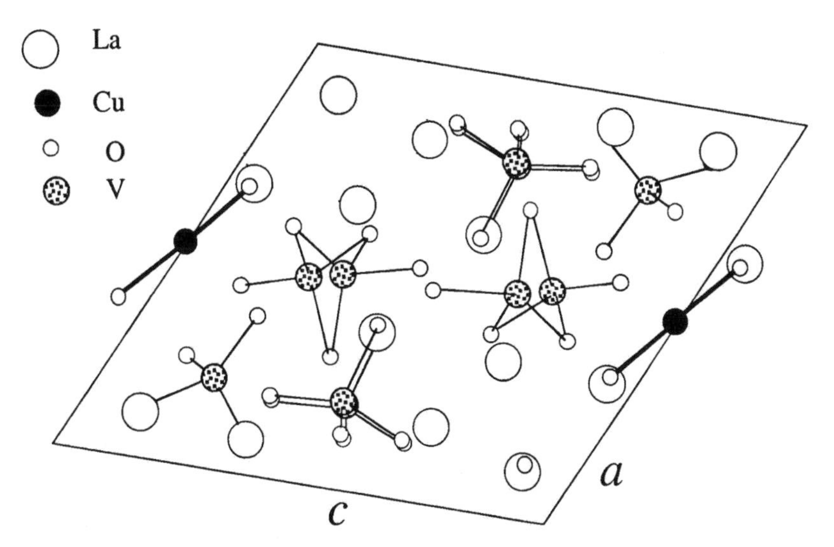

Figure 3. Projection of the $La_{14}V_6CuO_{36.5}$ structure viewed down the *b* axis. Some oxygen atoms and La-O bonds omitted for clarity.

Magnetic and electrical properties

The field cooled (FC) and zero field cooled (ZFC) susceptibilities of $La_{13}Cu_9V_4O_{38.5}$ are the same at all temperatures. The ZFC plot of the inverse molar susceptibility as a function of temperature is shown in Figure 4. There are three different linear regions in the inverse susceptibility plot indicating that there may be two magnetic transitions; one at 20 K and the other at 220 K. Below 20 K, θ is –8 K and the effective magnetic moment per Cu is 0.62 μ_B. For the linear region between 30 K $\leq$ T $\leq$ 220 K, $\theta = -107$ K and $\mu_{eff} = 1.28$ μ_B. For the third linear region above 280 K, θ is large (- 389 K) and the calculated μ_{eff} is 1.76 μ_B per Cu. This value is close to the expected spin-only moment of 1.73 μ_B for Cu^{2+} ($3d^9$: S = 1/2).

The magnetic behavior of $La_{13}Cu_9V_4O_{38.5}$ is strikingly similar to that of $La_4Cu_3MoO_{12}$, which contains Cu_3 trimers.[21,24] In the latter, antiferromagnetic spin interactions within the triangular clusters give rise to a net spin of S = 1/2 for each Cu_3 unit. The magnetic data of $La_{13}Cu_9V_4O_{38.5}$ shows a similar moment of 1.86 μ_B per Cu_3 trimer (3 × 0.62 μ_B) below 20 K; this moment is only slightly higher than the expected value for a S = 1/2 Cu_3 trimer. From the inverse susceptibility–temperature plot, it can be inferred that $La_{13}Cu_9V_4O_{38.5}$ would exhibit a true paramagnetic behavior above 389 K.

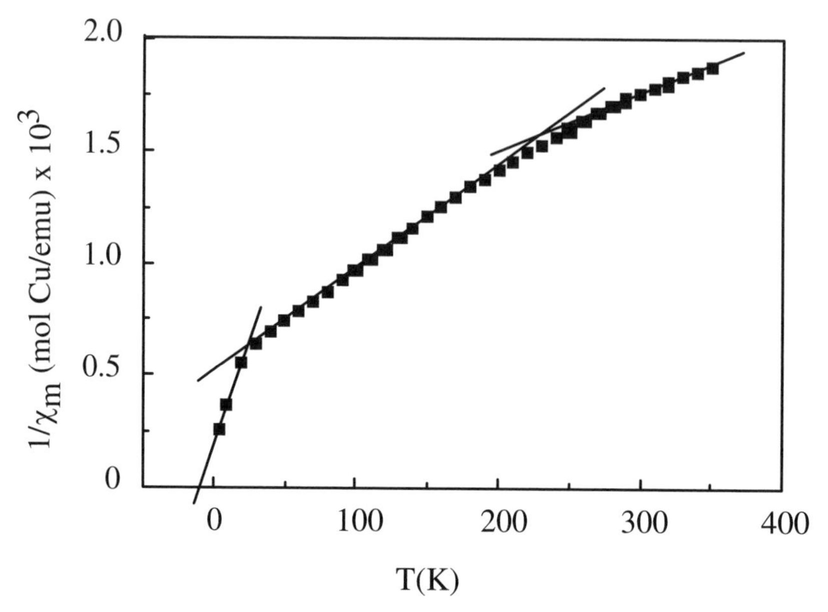

Figure 4. ZFC inverse magnetic susceptibility of $La_{13}Cu_9V_4O_{38.5}$ in the temperature range 5–300 K. Solid lines represent the three linear regions of the susceptibility.

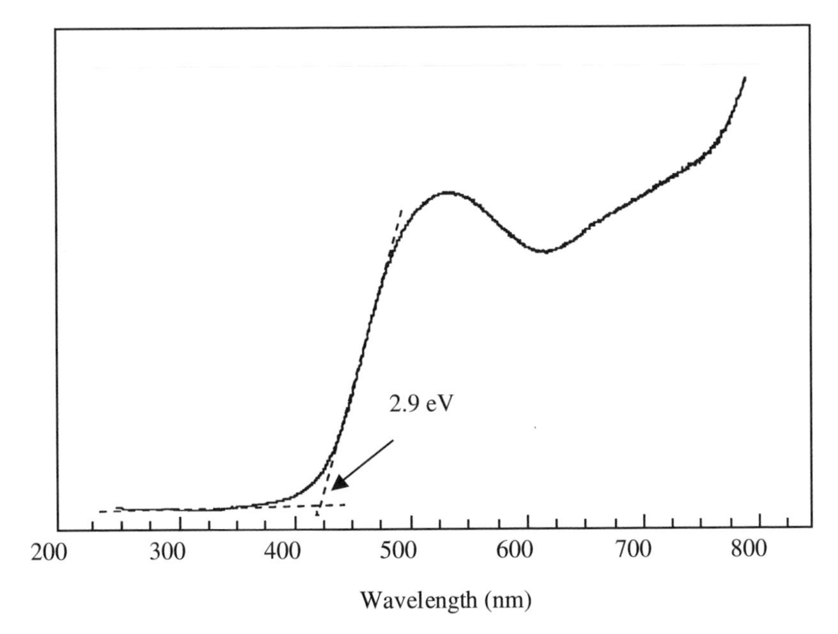

Figure 5. Diffuse reflectance spectrum of $La_{14}V_6CuO_{36.5}$. The dashed lines represent the estimation of the band gap from the band edge.

The corresponding temperature for $La_4Cu_3MoO_{12}$ is 460 K. Similar results were obtained by Poeppelmeier et al. for $La_3Cu_2VO_9$ [22,23].

Electrical resistivity measurements reveal that $La_{13}Cu_9V_4O_{38.5}$ is a semiconductor with room temperature resistivity of 4.2 ´ 10^5 Ohm.cm. The Arrhenius plot is linear between 200–300 K giving activation energy, 0.124 eV.

In order to survey the optical and transport properties of $La_{14}V_6CuO_{36.5}$, diffuse reflectance optical measurements were conducted on a polycrystalline sample. The reflectance spectrum in the visible region is shown in Figure 5. From a fit of the data between 380 and 500 nm, an optical band gap of ~2.9 eV is estimated for this compound. The material is transparent above ~420 nm although the tail of the band edge and the weak chromophore at 620 nm give the bulk polycrystalline samples a faint green hue. The optical band gap of the bulk material is smaller than those found in films of $CuAlO_2$ (3.5 eV) [1] and $CuScO_{2+x}$ (3.3 eV) [4].

DISCUSSION

The 1996 report of Jansson and co-workers [20] described a hexagonal $La_3Cu_2VO_9$ phase ($a_{Cu/V}$ = 14.4421, $c_{Cu/V}$ = 10.685 Å) with a large monoclinic supercell (a_m = 14.4427, b_m = 10.685, c_m = 52.07 Å and β = 106.1°). Comparison with Poeppelmeier's [21] hexagonal $La_4Cu_3MoO_{12}$ shows the relationship between the cells is $a_{Cu/V} = \sqrt{13}a_h$, $c_{Cu/V} = c_h$ where a_h and c_h are the hexagonal lattice parameters of the Cu/Mo phase. The striking similarities of the XRD patterns are also indicative of a very similar atomic ordering. Moreover, the magnetic susceptibility data for $La_{13}Cu_9V_4O_{38.5}$ is strongly suggestive of Cu_3 cluster formation as was observed for the $La_4Cu_3MoO_{12}$ phase [21,24]. Together, these data suggest an $La_4Cu_3MoO_{12}$–type structure for $La_{13}Cu_9V_4O_{38.5}$ containing discrete triangular Cu_3 clusters and Cu/V ordering. The structural refinements indicated the necessity of Cu/V ordering to form a 14.4 Å superstructure.

$La_{14}V_6CuO_{36.5}$ is a unique oxide containing isolated VO_4^{3-} tetrahedra and CuO_2^{3-} sticks. Seebeck effect measurements done on the thin films of $La_{14}V_6CuO_{36.5}$ indicated that the carriers were electrons and the compound was an n-type conductor regardless of the annealing conditions [25]. The lack of p-type conductivity in $La_{14}V_6CuO_{36.5}$ is presumably due to the low concentration of the Cu(I) centers relative to the Cu(I) p-type conductors.

ACKNOWLEDGEMENTS

I am grateful to the Link Foundation for the fellowship and the NSF-DMR for support of this work. I would like to thank my advisor Prof. Bryan W. Eichhorn for his valuable suggestions and support. I also would like to thank Prof. J. Gopalakrishnan and T. K. Mandal for the preliminary work on $La_3Cu_2VO_9$, Scott Sirchio for help with the WDS experiments, K.-S. Chang and Prof. Ichiro Takeuchi for the Seebeck effect measurements, Dr. James C. Fettinger for the single crystal analysis and Raymond E. Richardson for the diffuse reflectance measurement.

REFERENCES

(1) H. Kawazoe, M. Yasukawa, H. Hyodo, M. Kurita, H. Yanagi, H. Hosono, "P-type electrical conduction in transparent thin films of $CuAlO_2$" *Nature* 389, 939–942 (1997).

(2) A. F. Wells, *Structural Inorganic Chemistry*; 5th edition; Oxford University Press: New York, (1984).

(3) R. D. Shannon, D. B; Rogers, C. T. Prewitt, "Chemistry of noble metal oxides .1. Synthesis and properties of ABO_2 delafossite compounds" *Inorg. Chem.* 10, 713–716 (1971).

(4) N. Duan, A. W. Sleight, M. K. Jayaraj, J. Tate, "Transparent p-type conducting $CuScO_{2+x}$ films" *App. Phys. Lett.* 77,1325–1326 (2000).

(5) R. J. Cava, W. F. Peck, J. J. Krajewski, S. W. Cheong, H. Y. Hwang, "Electro-

chemical and high-pressure superoxygenation of $YCuO_{2+x}$ and $LaCuO_{2+x}$ delafossites" *J. Mater. Res.* 9, 314–317(1994).
(6) R. J. Cava, H. W. Zandbergen, A. P. Ramirez, H. Takagi, C. T. Chen, J. J. Krajewski, W. F. Peck, J. V. Waszczak, G. Meigs, R. S. Roth, L. F. Schneemeyer, "$LaCuO_{2.5+x}$ and $YCuO_{2.5+x}$ delafossites-materials with triangular Cu^{2+}delta planes" *J. Solid State Chem.* 104, 437–452 (1993).
(7) M. Shimode, M. Sasaki, K. Mukaida, "Synthesis of the delafossite-type $CuInO_2$" *J. Solid State Chem.* 151, 16–20 (2000).
(8) P. E. de Jongh, D. Vanmaekelbergh, J. J. Kelly, "Cu_2O: Electrodeposition and characterization" *Chem. Mater.* 11, 3512–3517 (1999).
(9) S.-G. Wang, W. H. E . Schwarz, "On closed-shell interactions, polar covalences, d shell holes, and direct images of orbitals: The case of cuprite" *Angew. Chem. Int. Ed.* 39, 1757–1758 (2000).
(10) R. J. Cava, "Structural chemistry and the local charge picture of copper-oxide superconductors" *Science* 247, 656–662 (1990).
(11) A. W. Sleight, "Chemistry of high-temperature superconductors" *Science* 242, 1519–1527 (1988).
(12) C. L. Teske, H. Muller-Buschbaum, "Alkaline earth metal oxocuprates, 4. $SrCu_2O_2$" *Z. Anorg. Allg. Chem.* 379, 113–& (1970).
(13) W. Carrillo-Cabrera, H. G. von Schnering, "Pentastrontium tris[tetraoxovanadate(V)] catena-monoxocuprate(I), $Sr_5(VO_4)_3(CuO)$—An apatite derivative with inserted linear (1)(infinity)[CuO](1–) chains" *Z. Anorg. Allg. Chem.* 625, 183–185 (1999).
(14) a) H. Klassen, R. Hoppe, "New oxocuprates(I)—on CsCuO" *Z. Anorg. Allg. Chem.* 497, 70–78 (1983). b) W. Losert, R. Hoppe, "On the $K_4[Ag_4O_4]$ relation" *Z. Anorg. Allg. Chem.* 524, 7–16 (1985).
(15) a) H. Klassen, R.Hoppe, "New oxocuprates(I)- on $Rb_3Cu_5O_4$" *Z. Anorg. Allg. Chem.* 494, 20–30 (1982). b) R. Hoppe, W. Losert, "New oxocuprates.I. on $Cs_3Cu_5O_4$, $Rb_2Kcu_5O_4$, $RbK_2Cu_5O_4$, and $K_3Cu_5O_4$" *Z. Anorg. Allg. Chem.* 504, 60–66 (1983).
(16) a) W. Losert, R. Hoppe, "The 1st ovocuprate(I) with dumb-bell-like anion-$KNa_2[CuO_2]$" *Z. Anorg. Allg. Chem.* 515, 87–94 (1984). b) R. Hoppe, W. Losert, "A new oxocuprate(I)—$K_3[CuO_2]$" *Z. Anorg. Allg. Chem.* 521, 69–78 (1985).
(17) P. Amann and A. Moller, "Synthesis, crystal structure, and reactivity of $Na_5[CuO_2](OH)_2$" *Z. Anorg. All. Chem.* 627, 2571–2575 (2001).
(18) H. Ohta, K. Kawamura, M. Orita, M. Hirano, N. Sarukura, H. Hosono, "Current injection emission from a transparent p-n junction composed of p-$SrCu_2O_2$/n-ZnO" *App. Phys. Lett.* 77, 475–477 (2000).
(19) A. Kudo, H. Yanagi, K. Ueda, H. Hosono, H. Kawazoe, Y. Yano, "Fabrication of transparent p-n heterojunction thin film diodes based entirely on oxide semiconductors" *App. Phys. Lett.* 75, 2851–2853 (1999).
(20) K. Jansson, I. Bryntse, Y. Teraoka, "Synthesis of a new compound in the La-Cu-V-O system: $La_3Cu_2VO_9$" *Mat. Res. Bull.* 31, 827–835 (1996).
(21) D. A. Vander Griend, S. Boudin, V. Caignaert, K. R. Poeppelmeier, C. Wang, V. P. Dravid, M. Azuma, M. Takano, Z. Hu, J. D. Jorgensen, "$La_4Cu_3MoO_{12}$: A novel cuprate with unusual magnetism" *J. Am. Chem. Soc.* 121, 4787–4792 (1999).
(22) S. Malo, D. A. Vander Griend, K. R. Poeppelmeier, Y. Wang, V. P. Dravid, "Crys-

tal symmetry of $La_3Cu_2VO_9$ and $La_4Cu_3MoO_{12}$ derived from the $YAlO_3$ hexagonal structure by transmission electron microscopy" *Solid State Sci.* 3, 17–23 (2001).
(23) D. A. Vander Griend, S. Malo, K. R. Poeppelmeier, V. P. Dravid, "$La_3Cu_2VO_9$: A surprising variation on the $YAlO_3$ structure-type with 2D copper clusters of embedded triangles" *Solid State Sci.* 3, 569–579 (2001).
(24) M. Azuma, T. Odaka, M. Takano, D. A. V. Griend, K. R. Poeppelmeier, Y. Narumi, K. Kindo, Y. Mizuno, S. Maekawa, "Antiferromagnetic ordering of S=1/2 triangles in $La_4Cu_3MoO_{12}$" *Phys. Rev. B* 62, R3588–R3591 (2000).
(25) The work on $La_{14}V_6CuO_{36.5}$ is in press, *J. Mater. Chem.* 12, 3839–3842 (2002).

Engineering Hybrid Nanostructured Composites as Active Materials for Lithium Rechargeable Batteries

Huan Huang

Chemistry Department, University of Waterloo

200 University Avenue West

Waterloo, Ontario N2L 3G1, Canada

Research Advisor: Dr. Linda F. Nazar

ABSTRACT

In this paper, we present two nanostructured composites as electroactive materials for lithium ion batteries. Nanocomposite V_2O_5/C-PEG is composed of V_2O_5 xerogel coating on polymer electrolyte grafted-carbon black. Owing to the facile transports of electrons and Li^+ ions involved in the unique nanostructure, this nanocomposite exhibits greatly improved electrochemical performance. In particular, its excellent rate sustainability makes this composite material very promising for high power/current applications. Nanocomposite $LiFePO_4$/C is made up of nanocrystalline $LiFePO_4$ embedded in a conductive porous carbon framework. This composite can successfully reach the full capacity at a mediate rate at room temperature with good rate capability and excellent electrochemical stability. Both nanostructures are versatile and applicable to other active materials.

INTRODUCTION

New applications, such as portable electronic devices and electric vehicles, demand advanced energy storage systems. Among these systems, the lithium ion rechargeable battery is one of the superior alternatives, owing to its high voltage output, high specific energy, energy density, and satisfactory cycling life [1]. However, accessing high charging and discharging rate without losing capacity is one of the biggest challenges faced in the lithium battery [1]. This is due to the often sluggish kinetics of Li^+ ion transport in solid electrodes, which hinders it from high power/current applications.

Another challenge in this field is to overcome the insulating nature of transition metal polyanionic compounds in order to access their high capacities [2]. Among this class of compounds, olivine-type $LiFePO_4$ is of greatest interest [2–4], because of its large theoretical capacity (170 mAh/g), moderately high working potential (~3.5 V), and excellent electrochemical cycling stability. In addition this material is naturally abundant, inexpensive and environmental benign. Unfortunately, only partial lithium was reported to be able to reversibly extracted/inserted in $LiFePO_4$ due to its very poor electronic conductivity. For example, at a moderate current, 23 mA/g (C/7), the available capacity is only 70 mA/g (~0.4 Li) [5].

Aiming at the above two goals, through anostructured composite approach we proposed two nanocomposite [6, 7]. One is coating the active metal oxide V_2O_5 on polymer electrolyte-modified carbon, that is, V_2O_5/C-PEG (Structure I in Figure 1), and another is incorporating electroactive crystalline phosphates with a web-like carbon xerogel framework, such as $LiFePO_4$/C (Structure II in Figure 1).

The anticipated advantages of structure I arising from this hybrid combination are synergistic. The carbon black will favor electronic conductivity and the stability of the coated active material; likewise, the polymer electrolyte grafted on the carbon surface and interspersed between the host material layers will improve the ionic conductivity and thus ease the diffusion of Li^+ ions, and the mechanical strength of this composite will be reinforced by the polymer chains. Moreover, since the coating of active materials is very thin (the thickness is in the magnitude of several tens or several nanometers), the Li^+ diffusion path will be greatly reduced.

In structure II, carbon xerogel was chosen as the conductive framework, because of its highly intercross-linked microstructure, relatively high electronic conductivity (10–100 S/cm), and excellent intrinsic adhering ability as well. Anticipated advantages of this nanocomposite include: web-like carbon xerogel framework interconnected throughout the active material particle can dramatically improve the electronic conductivity of the particle, which will not only ensure good current collection efficiency, but also

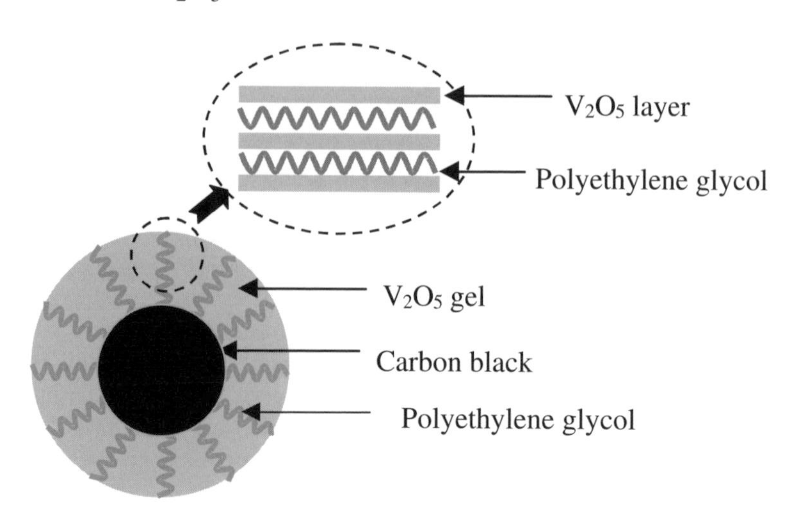

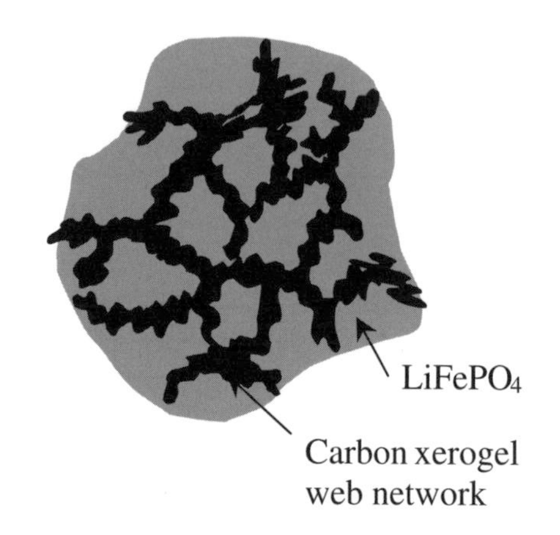

Figure 1. Schematic illustration of the two designed nanostructures. Structure I: V_2O_5 gel coated polymer electrolyte grafted-carbon black (V_2O_5/C-PEG). The inset indicates that the polymer chains are inserted between V_2O_5 layers; Structure II: $LiFePO_4$/C composite made by incorporating $LiFePO_4$ with web-like carbon xerogel framework.

make the inner part of the particle to be electron readily accessible; web-like carbon xerogel framework can restrain the crystalline $LiFePO_4$ active material from growth during its formation at high temperature, which will significantly reduce the actual Li^+ ion diffusion distance; the active material is hold by the carbon web framework, making the particle mechanically strong, and thus favoring the stability.

Here we report the synthesis methods of V_2O_5/C-PEG and $LiFePO_4$/C nanocomposites and their greatly improved electrochemical properties including the capacity, the rate capability and the cycling stability.

EXPERIMENTAL

Preparation of V_2O_5/C-PEG hybrid nanocomposite

Our procedure for coating active materials on polymer electrolyte-grafted carbon blacks is illustrated in Scheme I. Acetylene black was used as the core support, because of its well-known high electronic conductivity (~10 S/cm) and small particle size (40–50 nm). It was treated with concentrated HNO_3/H_2SO_4 (1:1 in volume) acids at 70 for 12 hours, giving rise to the formation of surface functional groups [8]. IR spectra show a new band appeared at 1716 cm^{-1}, which is ascribed to the carboxylic group (-COOH), the key reaction group for polymer grafting.

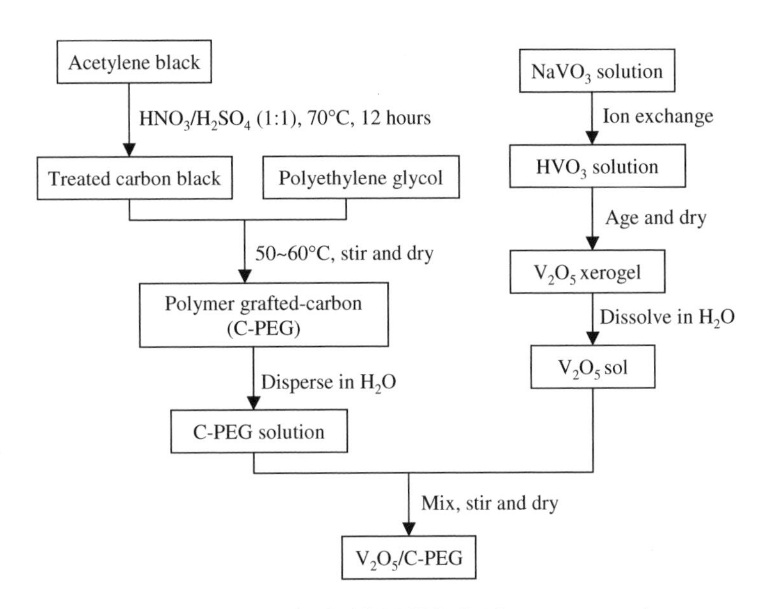

Scheme I. Procedure to prepare V_2O_5/C-PEG hybrid nanocomposite.

To grow polyethylene glycol (PEG, MW: 4600, Aldrich, 98%) on carbon surface, the acid-oxidized carbon black was dispersed in water and then mixed with a PEG solution. The solution was heated at 50–60 and stirred continuously, while the carboxylic group on carbon surface reacted with the hydroxyl group of the polymer so that polymer could be chemically attached [9]. After evaporation of water, PEG grafted-carbon black (C-PEG) was obtained. The weight ratio of carbon and polymer was 2 to 1.

V_2O_5 xerogel was prepared by protonation of a metavanadate solution through the ion exchange method [10]. A 1.0 M $NaVO_3$ (Aldrich, 98%) solution was passed through an ion exchange column containing Dowex 50WX2–100 cation-exchange resin, and a yellowish HVO_3 solution was collected. Upon standing, HVO_3 polymerized within several hours to form a red V_2O_5 sol, which thickened and became a homogeneous red gel after several days. Freely standing V_2O_5 xerogel thin films were formed on polyethylene plates after evaporation of water.

V_2O_5 gel was completely dissolved in deionized water with the aid of ultrasound and produced a dark red clear solution, in which a solution containing well-dispersed C-PEG was added. The mixed solution was magnetically stirred at 50~60 to evaporate water, and V_2O_5/C-PEG nanocomposite powder was produced when dried. The weight ratio of V_2O_5 and C-PEG was 4:1.

In comparison with this novel designed V_2O_5/C-PEG nanocomposite, bulk V_2O_5 xerogel and V_2O_5 gel coated directly on carbon black (V_2O_5/C) were prepared.

Preparation of $LiFePO_4$/C nanocomposite

Scheme II gives the procedure to prepare $LiFePO_4$/C nanocomposite [11]. In the present work, resorcinol (Aldrich, 98%) and formadehyde (Aldrich, 37 wt.%) with R/F mole ratio of 1:2 were dissolved in deionized water with sodium carbonate (Na_2CO_3, Aldrich, 99.5%) added as the basic catalyst for the polymerization. The pH of the solution was adjusted to 6 6.5. The solution was sealed in a bottle and placed in an oven previously set at 85 C. Typically, gelation occurred in several hours and the solution was cured for 3~5 days to obtain dark red RF gel. The cooled gel was immersed in acetone (EM, 99.5%) to exchange out the water.

Stoichommetric amount of the precursors, $LiCH_3COO$ (Aldrich, 98%), $Fe(CH_3COO)_2$ (Alfa, 29%-33% based on Fe) and $NH_4H_2PO_4$ (Aldrich, 99%) were mixed completely with RF gel by ball milling at 550 rpm for 2 hours. After dried at 80 C in an oven, the mixture was transferred in a tube furnace under N_2 flow. The furnace was first heated to 350 C for 5 hours to decompose the precursors, and then consequently heated to 700 C for 10 hours. At this high temperature, carbon xerogel was produced through pyrolysis of RF gel, while $LiFePO_4$ was formed from its precursors *in-situ*. The furnace was cooled to room temperature to obtain $LiFePO_4$/C

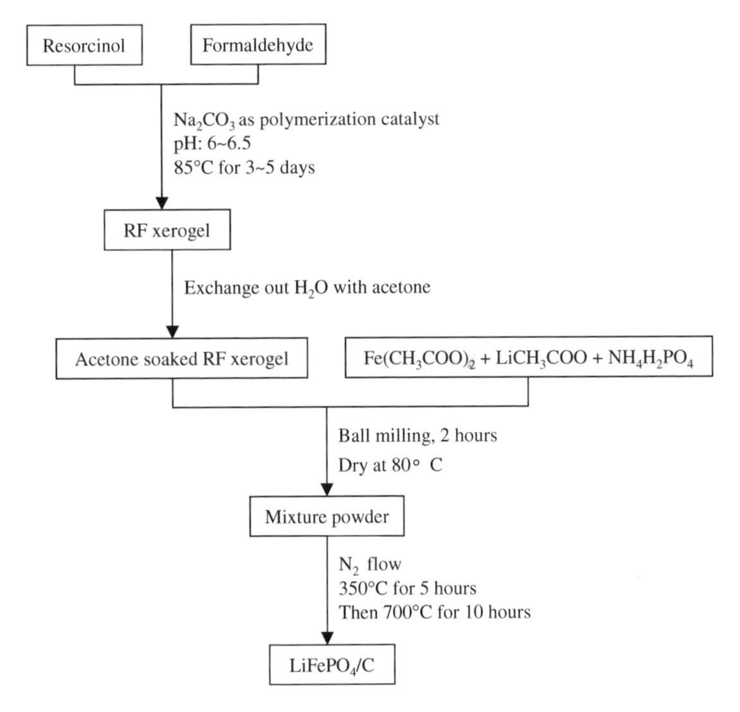

Scheme II. Procedure to prepare $LiFePO_4$/C nanocomposite.

nanocomposite. Carbon content in the composite is about 15wt%. For comparison, bulk $LiFePO_4$ was made by solid-state method.

Electrochemical measurement

Electrochemical cells were constructed using a Swagelok™ design, utilizing 1M $LiPF_6$/EC-DMC (Merck) as the electrolyte. For V_2O_5/C-PEG, a cathode containing nanocomposite material, Super S, and PVDF with the weight ratio 80:10:5 were used, and for $LiFePO_4$/C nanocomposites, cathodes comprised of the composite, Super S (MMM), and PVDF binder with 90:5:5. Cells were assembled in a glove box under Ar with O_2 and H_2O lower than 5 ppm.

RESULTS AND DISCUSSION

Electrochemical performance of V_2O_5/C-PEG hybrid nanocomposite

The discharge/charge profile of the first cycle

The electrochemical response of our hybrid nanostructured material V_2O_5/C-PEG is greatly enhanced compared to a standard V_2O_5 xerogel electrode, as seen in Figure 2. All experiments were performed under the same conditions with a current density of 10 mA/g and a voltage window of 3.8–1.8 V. Data on the V_2O_5 xerogel are comparable to results previously reported [12]. Less than theoretical capacity is observed for the xerogel (1.3Li/V_2O_5 or 182 mAh/g for the discharge, and 1.1Li/V_2O_5 or 154 mAh/g for the charge) and some significant irreversibility (Figure 2 (a)). For V_2O_5/C-PEG composites almost complete reversibility is achieved, along with significantly reduced polarization and an almost doubled capacity (2.26 Li/V_2O_5 or 320 mAh/g, Figure 2 (c)). These improvements are attributed to a short Li^+ diffusion distance and an increase of electronic contact upon employing this hybrid nanostructure. Note that better response is also seen for the V_2O_5 directly coated on the carbon black particle (V_2O_5/C, Figure 2 (b)), which has a reversible capacity of 240 mAh/g, but it is not as significant as the nanocomposite with the polymer electrolyte present.

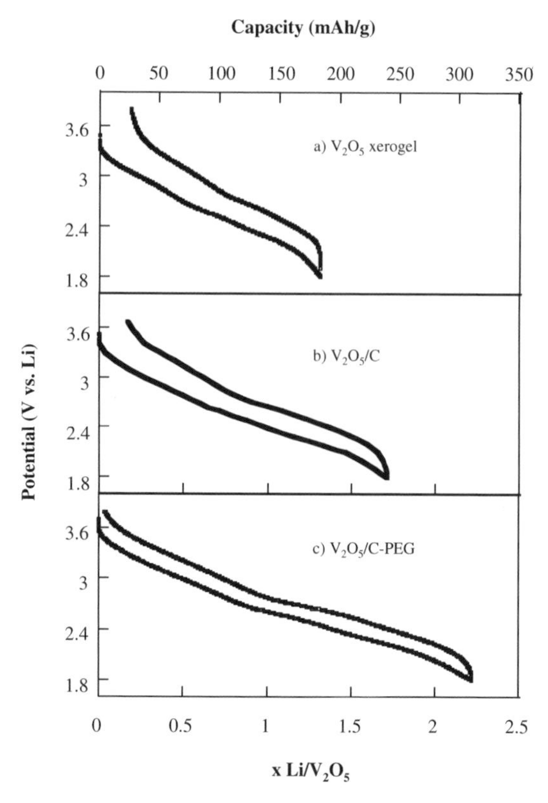

Figure 2. The first discharge/charge profiles of (a) V_2O_5 xerogel, (b) V_2O_5/C, and (c) V_2O_5/C-PEG.

Rate behavior

The rapid ionic and electronic transports in the hybrid nanostructured material are confirmed by varying the applied current (both for charging and discharging) to probe the comparative kinetic response. The effect of the current density on the working capacity on these three materials, up to current rates as high as 2000 mA/g (corresponding to a discharge rate of 14C—discharge or charge in 0.07 hours) are shown in Figure 3.

The capacity of the V_2O_5 xerogel (Figure 3 (a)) decreased rapidly when a higher current was applied. At a low current of 10 mA/g, the reversible capacity for the gel was 160 mAh/g; increasing the current to 500 mA/g (~3.5C) curtailed the available capacity to only *ca.* 30 mAh/g, which is less than 20% of the initial value at the lower rate. This poor kinetic response is typical of a combination of electronic and ionic factors that limit redox accessibility under rapid rate conditions. Some improvement in the kinetic response is visible with V_2O_5/C (Figure 3 (b)) as a result of the intimate contact of the oxide with the carbon. For example, at 500 mA/g, it main-

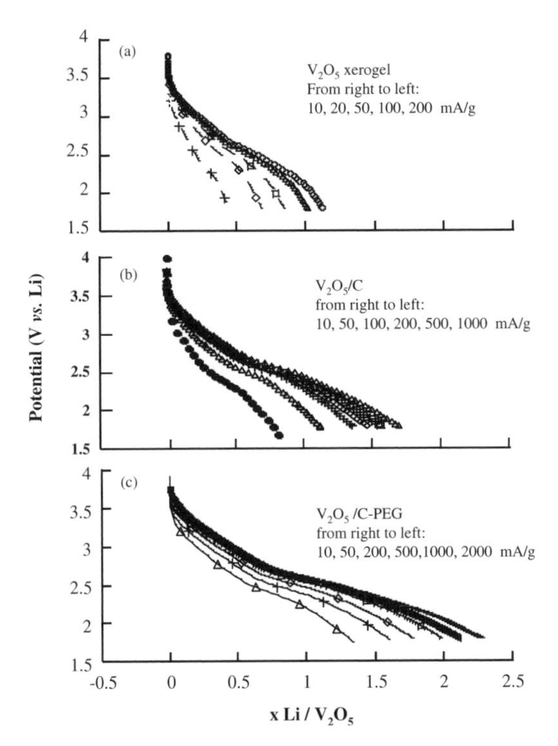

Figure 3. Rate performance of (a) V_2O_5 xerogel, (b) V_2O_5/C, and (c) V_2O_5/C-PEG.

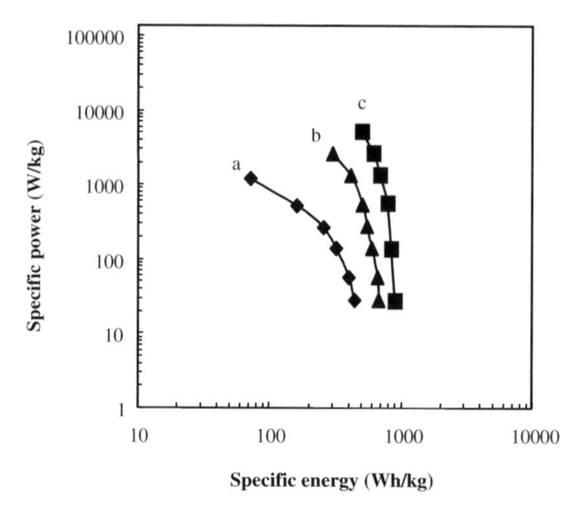

Figure 4. Ragone plots of (a) V_2O_5 xerogel, (b) V_2O_5/C, and (c) V_2O_5/C-PEG.

tained a capacity of 160 mAh/g, 66% of the capacity (242 mAh/g) at 10 mA/g. For the novel nanostructured composite, however, the rate capability was greatly improved. V_2O_5/C-PEG sustained current densities as high as 200 fold greater than the xerogel while maintaining an even higher capacity than the initial capacity of V_2O_5 gel at a low rate. At extremely high current densities of 1000 (7C) and 2000 mA/g (14C), capacities of 225 and 190 mAh/g were obtained respectively for V_2O_5/C-PEG that correspond to about 71% and 60% of the capacity at a low current density. Based on its excellent kinetic response, this new nanostructured composite is very suitable for high current/power applications. This is summarized in the Ragone plot, which is specific power *versus* specific energy (Figure 4). The curve for the xerogel levels off at high power, indicating that the available energy becomes less rapidly. V_2O_5/C-PEG exhibits much better high power output capability. Even at 5000 W/kg, 70% of the total specific energy can be used.

Cycling stability

The comparison in Figure 5 with the other composite materials indicates that the hybrid nanostructured material has not only substantially higher sustainable capacities, but also better cycling stability. Note the almost complete reversibility does not lead to a drop in capacity in the initial few cycles, whereas the other two materials display a relatively common—initial capacity decrease. This behavior is usually ascribed to irreversible reaction of lithium with surface defects, or entrapment in the struc-

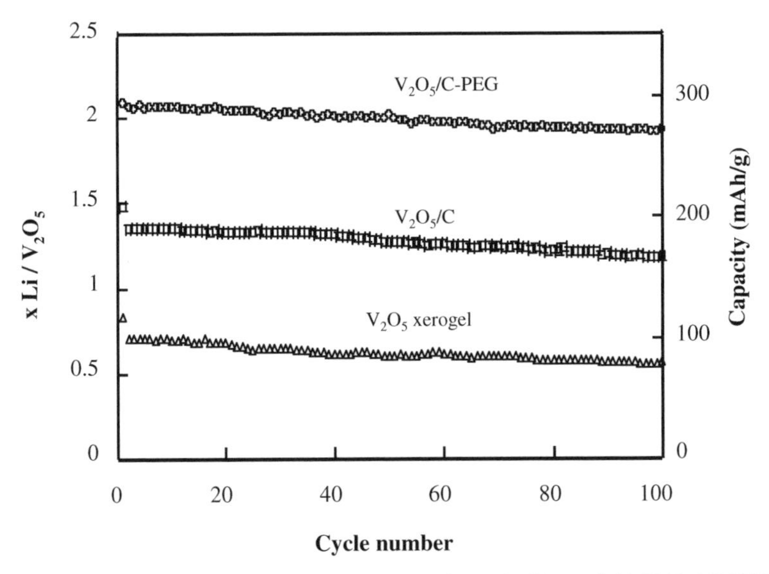

Figure 5. Cycling stability of (a) V_2O_5 xerogel, (b) V_2O_5/C, and (c) V_2O_5/C-PEG.

ture, which clearly does not occur in the nanostructured material. After 100 cycles at 100 mA/g, V_2O_5/C-PEG nanocomposite can maintain nearly 90% of its initial capacity, and V_2O_5/C can maintain 82%, while V_2O_5 gel only 78%.

Electrochemical performance of $LiFePO_4$/C nanocomposite

Charge/discharge behavior of $LiFePO_4$/C

The galvanostatic charge/discharge profile of the first cycle on the electrode using nanocomposite $LiFePO_4$/C is shown as curve a (solid line) in Figure 6, together with that on the electrode from bulk $LiFePO_4$ as curve b (dash line) for comparison. Both electrodes were operated under similar conditions with the charge/discharge rate of C/10 and the potential window 2.8 4.0 V.

In curve b, only 0.46 Li/$LiFePO_4$ is extracted during charge, and 0.34 Li/$LiFePO_4$ is reinserted during discharge, and no charge or discharge plateau was evident. The reversible capacity is as low as 60 mAh/g and irreversibility is as high as 26%, similar to the previously reported results performed at a close rate (C/7) [5]. The low capacity and high irreversibility are mainly ascribed to its very low electronic conductivity ($< 10^{-7}$ S/cm) and large particle/grain size, which lead to part of active material is electronic insulating and inaccessible by Li^+ ions.

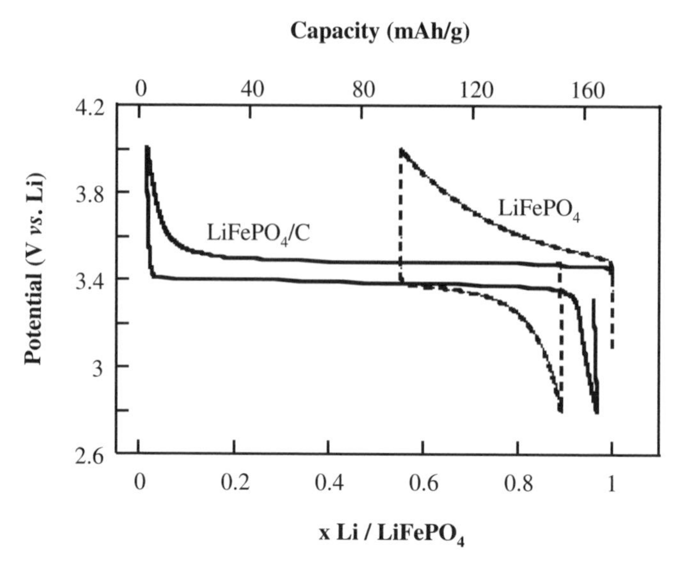

Figure 6. The first charge/discharge of (a) bulk $LiFePO_4$ (dash line) and (b) the $LiFePO_4$/C composite (solid line).

By employing our designed nanostructured composite, a different charge/discharge feature was observed and great improvement was achieved. In curve a, on charging the potential is quickly raised to a flat plateau at ~3.5 V, and nearly all lithium (0.983 Li/$LiFePO_4$) is extracted when the potential reaches the upper limit. On discharging, the potential drops quickly to a plateau at ~3.4 V, and 0.952 Li/$LiFePO_4$, corresponding to a capacity of 162 mAh/g, was able to recover. Compared to bulk $LiFePO_4$, therefore, the composite material has displayed not only the greatly improved capacity, but also the much lower irreversibility loss and smaller polarization.

The reversibility is also evident from their incremental capacity voltammagrams, which are derived from curve a and b in Figure 6 and are plotted in Figure 7 as curve a and b, respectively. $LiFePO_4$/C composite shows an oxidation peak at 3.48 V and a reduction peak at 3.39 V. The shape of both peaks is well-defined sharp peak and the peak separation is very narrow of ~90 mV, indicating an excellent reversible redox system. While pure $LiFePO_4$ displays broad oxidation and reduction peaks at 3.52 and 3.36 V respectively with a wider peak separation of ~160 mV, and thus much worse reversibility.

The greatly improved properties in capacity, reversibility and polarization of our nanocomposite material, compared to bulk $LiFePO_4$, are ben-

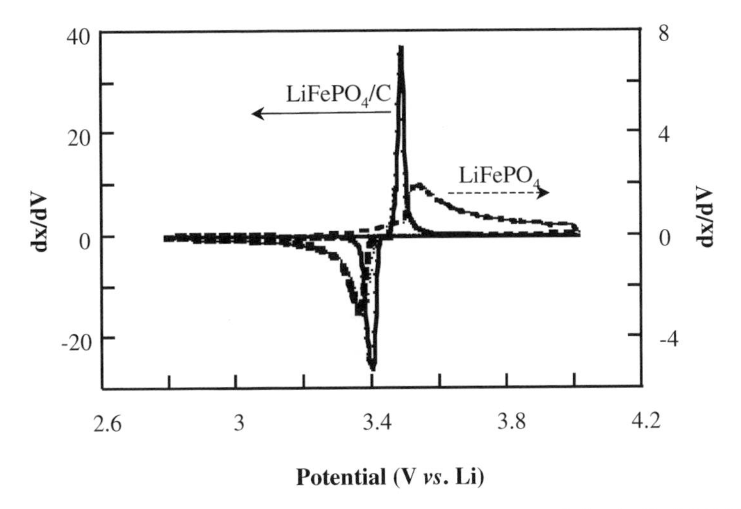

Figure 7. Plots of the incremental capacity *versus* potential for bulk (a) $LiFePO_4$ (dash line) and (b) the $LiFePO_4$/C composite (solid line).

efited from the designed hybrid nanostructure, that is, the active materials is evenly incorporated throughout electronically conductive carbon web, which in return provides good electronic contact and facile electron transport path, and enhances current collect efficiency.

Rate performance of $LiFePO_4$/C nanocomposite

As described above, our nanocomposite material can successfully reach the full capacity of $LiFePO_4$ at C/10, a moderate rate. However, in some new applications, higher charge/discharge rates are necessary. To examine the rate performance of this nanocomposite, the cell was charged/discharged at various rates including C/10, C/5, C/2, C, 2C, and 5C. Figure 8 records the discharge curves

Generally as the rate increases, the reversible lithium amount or the capacity decreases and the polarization increases. At a low rate, such as C/5 (i.e. 32.5 mA/g), a capacity of 157 mAh/g is obtained indicating over 90% of Li is accessible. When the rate increases to a medium rate, e.g. C/2 (81.4 mA/g), available capacity is 150 mAh/g is, close to 90% of the full capacity. When the rate increases further to a high rate of 1 C (163 mA/g), 83% of the total capacity, that is, 142 mAh/g is resulted. Even at a very fast rate 5C, ~70% Li was still reversible, reaching a capacity of 119 mAh/g.

For a poor conductive phosphate material, the results on our nanocomposite demonstrated a remarkable improvement of the rate re-

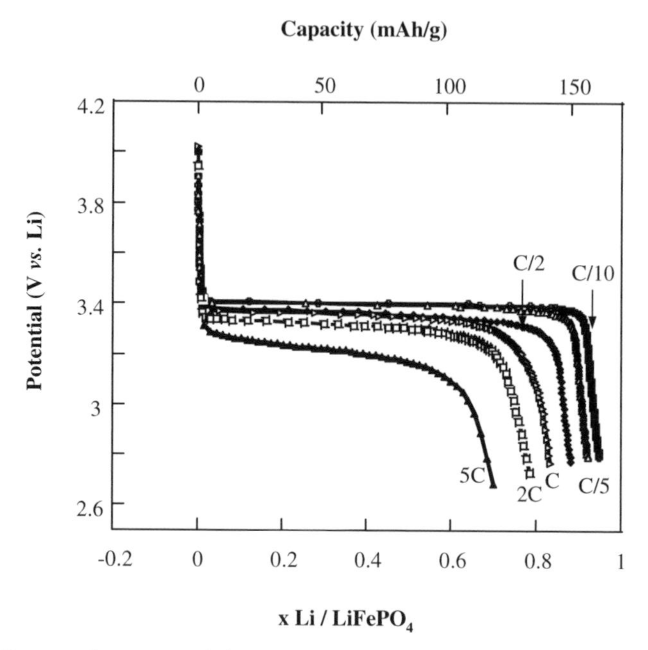

Figure 8. Rate performance of the $LiFePO_4$/C composite at room temperature.

tention. This rate capability is comparable to that of a commercial cell using $LiCoO_2$–based cathode, which normally can maintain 80% of the total capacity at 1 C rate, although $LiCoO_2$ has fast kinetics compared to bulk $LiFePO_4$. The rate performance of our nanocomposite meets the requirement for high rate/power applications.

Cycling stability

The cycling stability on bulk $LiFePO_4$ has been claimed to be excellent, owing to its strong framework structure and very low disturbing on the volume upon insertion/extraction. Using the nanocomposite material, we studied its cycling stability throughout at various rates: a medium rate of C/5 for 100 cycles, a fast rate of C/2 for 200 cycles, and a very fast rate of 5C for 800 cycles, shown as (a), (b), and (c) in Figure 9, respectively. At C/5, subtle capacity loss (< 4%) was observed over 100 cycles. At C/2, a slightly increase of capacity from 150 mAh/g to 154 mAh/g was observed in initial cycles, followed by a very small capacity loss lower than 5% at 200th cycle compared to the maximum capacity. At 5C, the initial increase of capacity is more pronounced, from 106 to 120 mAh/g, and the loss of capac-

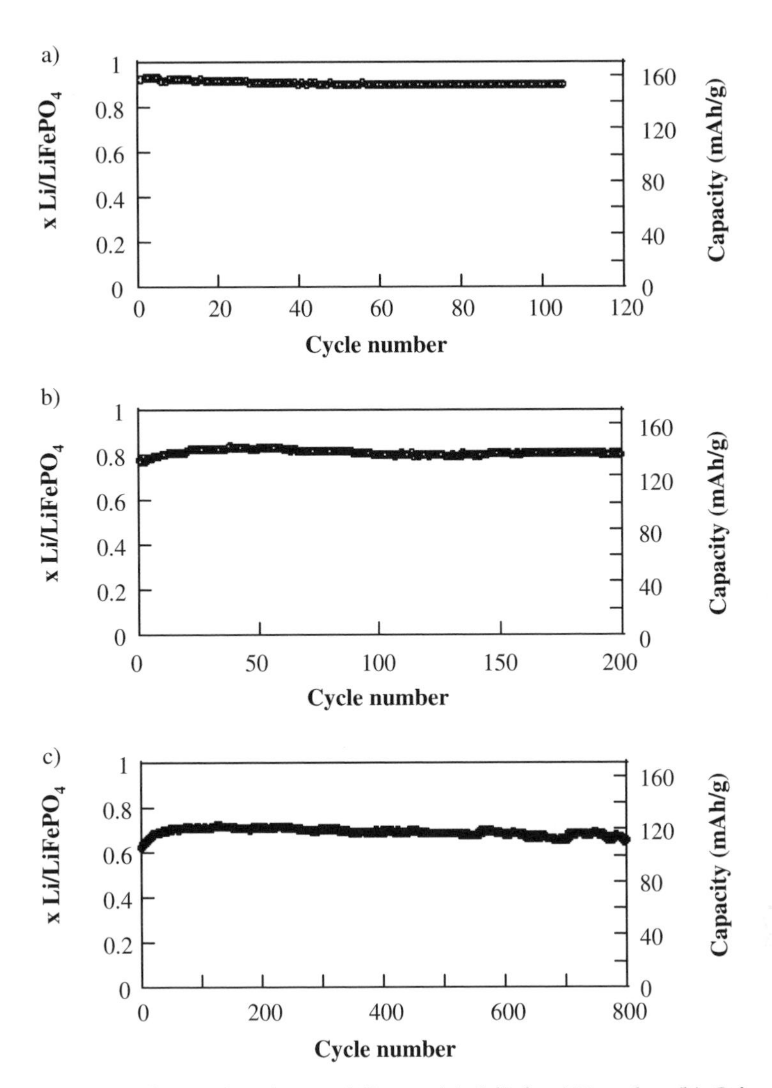

Figure 9. Electrochemical cycling stability at (a) C/5 for 100 cycles, (b) C for 200 cycles, and (c) 5C for 800 cycles.

ity is only 8% after 800 cycles. The composite electrodes have displayed excellent cycling stability under various all tested rates, making this material extremely suitable for long duration applications, such as the power for the electrical vehicle, which requires more than 1000 cycles or the stationary power source, which needs to run over 10 years.

CONCLUSION

Hybrid nanostructured V_2O_5/C-PEG can be classified as fast transport materials on the basis of their kinetic response to high current densities. This is the result arising from multiple contributions including facile Li^+ transport owing to the polyelectrolyte component, electron transport due to the coating of the V_2O_5 component to the conductive carbon surface, and the short diffusion path length. One of successful nanocomposites displayed a capacity as high as 320 mAh/g, and outstanding rate sustainability, which is capable as a supercapacitor.

Nanocomposite $LiFePO_4$/C is prepared by *in-situ* formation of $LiFePO_4$ and carbon xerogel, which ensures particle size minimization and intimate carbon contact, and thus optimizing accessibility of the redox centers. Such composite successfully achieves the full capacity at a moderate rate and large capacity at high rates with excellent cycling stability, making this material an almost ideal cathode.

ACKNOWLEDGEMENTS

This work was financially supported by the NSERC through funding from Strategic and Research Grants. I am grateful to the Link Foundation Energy Fellowship and Ontario Graduate Scholarship for their generous support. Special appreciation must be given to my advisor Prof. Linda F. Nazar for her invaluable guidance and support.

REFERENCES

[1] R. J. Brod, "Recent Developments in Batteries for Portable Consumer Electronics Applications", *Electrochem. Soc. Interface,* 8(3), 20–23 (1999)

[2] J. M. Tarascon and M. Armand, "Issues and Challenges Facing Rechargeable Lithium Batteries", *Nature,* 414, 359–367 (2001)

[3] A.K. Padhi, K.S. Nanjndaswamy, and J.B. Goodenough, "Phospho-olivines as Positive-Electrode Materials for Rechargeable Lithium Batteries", *J. Electrochem. Soc.,* 144 (4), 1188–1194 (1997)

[4] A. Yamada, S. Chung, and K. Hinokuma, "Optimized $LiFePO_4$ for Lithium Battery Cathodes", *J. Electrochem. Soc.,* 148, A224–A229 (2001)

[5] A. S. Andersson, J. O. Thomas, B. Kalska, and L. Haggstrom, "Thermal Stability of $LiFePO_4$–Based Cathodes", *Electrochem. Solid-State Lett.,* 3(2), 66–68 (2000)

[6] H. Huang and L. F. Nazar, "Grafted Metal Oxide/Polymer/Carbon Nanostructures Exhibiting Fast Transport Properties", *Angew. Chem. Intl. Ed.,* 40 (20), 3880–3884 (2001)

[7] H. Huang, S.-C. Yin, and L. F. Nazar, "Approaching Theoretical Capacity of

$LiFePO_4$ at Room Temperature at High Rates", *Electrochem. Solid-State Lett.*, 4(10), A170–A172 (2001)
[8] *Carbon black: science and technology*, Eds: J-B Donnet, R. C. Bansal, and M-J. Wang, New York: Dekker (1993)
[9] J-H Lin, H-W Chen, K-T Wang, and F-H Liaw, "A novel method for grafting polymers on carbon blacks", *J. Mater. Chem.*, 8, 2169–2173 (1998)
[10] J. Livage, "Vanadium Pentoxide gels", *Chem. Mater.*, 3 (4), 578–593 (1991)
[11] C. Lin and J. A. Ritter, "Carbonization and activation of sol–gel derived carbon xerogels", *Carbon*, 38(6), 849–861 (2000)
[12] M. Lira-Cantu and P. Gomez-Romero, "The Organic-Inorganic Polyaniline/V_2O_5 System. Application as a High-Capacity Hybrid Cathode for Rechargeable Lithium Batteries", *J. Electrochem. Soc.*, 146(6), 2029–2033 (1999)

A Two-Electron Platinum Reagent

Hershel Jude

Department of Chemistry
University of Cincinnati
P.O. Box 210172
Cincinnati, OH 45221–0172
Research Advisor: Dr. William B. Connick

ABSTRACT

The compounds [Pt(tpy)(pip$_2$NCN)](BF$_4$) (**1**(BF$_4^-$)), [Pt(tpy)(pip$_2$NCNH$_2$)](PF$_6$)$_3$ (**2**(PF$_6^-$)$_3$) and [Pt(tpy)(dmph)](BF$_4$) (**3**(BF$_4^-$)) (tpy = 2,2':6',2"-terpyridine, pip$_2$NCN$^-$ = 2,6–(CH$_2$N(CH$_2$)$_5$)$_2$–C$_6$H$_3^-$) have been synthesized and characterized by elemental analysis, ^{1}H NMR spectroscopy and mass spectrometry. The structure of the chloride salt of **2** was confirmed in a preliminary X-ray crystallographic study. In all three square planar platinum(II) complexes, the tpy ligand is tridentate. In compounds **1** and **2**, the piperdyl ligand is monodentate, bonded to Pt(II) through the phenyl ring. The cyclic voltammagram of Pt(tpy)(pip$_2$NCN)$^+$ (**1**) exhibits two reversible one-electron terpyridine centered reductions in acetonitrile at E°'=–0.98 V (ΔE$_p$=65 mV, i_{pc}/i_{pa}=0.80) and E°'= 1.50 V (ΔE$_p$=61 mV, i_{pc}/i_{pa}=0.52) vs. Ag/AgCl at 0.25 V/s. In contrast to other platinum(II) complexes, including **2** and **3**, this compound undergoes a nearly reversible two-electron platinum centered oxidation at 0.40 V (ΔE$_p$=43 mV at 0.01 V/s).

INTRODUCTION

The development of efficient multi-electron photocatalysts that utilize sunlight to convert substrates to chemical fuels poses a significant challenge. One approach focuses on transition metal-based complexes capable of undergoing photoinduced outer-sphere electron-transfer reactions. The premise is that the enhanced reactivity of an excited complex can be harnessed in electron-transfer reactions that yield oxidized and reduced products. Because many important target reactions (*e.g.*, splitting water into H_2 and O_2) require multiple electron-transfer steps, there is interest in developing photo-reagents capable of undergoing multi-electron transfer reactions [1].

Platinum complexes are intriguing candidates for investigation because of their tendency to undergo two-electron changes in oxidation state. For example, Pt(II) and Pt(IV) complexes are stable, but monomeric Pt(III) complexes tend to be unstable. A serious drawback is that the interconversion between these oxidation states by outer-sphere electron transfer is slow because of the accompanying large molecular reorganization [2]. Pt(II) complexes with d^8 electron valence configurations favor a four-coordinate square planar geometry, whereas d^6 electron Pt(IV) complexes favor a six-coordinate octahedral geometry. The accompanying ligand association and/or dissociation steps cause these two-electron reactions to be electrochemically irreversible with large overpotentials. This lack of reversibility is a critical problem because it prevents regeneration of the active chromophore and completion of the catalytic cycle.

In order to facilitate outer-sphere two-electron transfer reactions of Pt(II) and Pt(IV) complexes, we are designing compounds with ligand systems capable of stabilizing both square planar and octahedral coordination geometries. The strategy forming the basis of the current investigation involves binding of two potentially meridional coordinating ligands to a platinum center. In the square planar Pt(II) case, one ligand is tridentate and the second is monodentate, coordinating through the central binding site. Upon transfer of two electrons the pendant groups of the monodentate ligand are expected to bind to the metal, stabilizing the oxidized Pt(IV) center (Figure 1).

The platinum(II) compound pictured in Figure 2, $Pt(tpy)(pip_2NCN)^+$ (**1**) (tpy=2,2':6',2"-terpyridine, pip_2NCN^-=2,6–$(CH_2N(CH_2)_5)_2$–$C_6H_3^-$), is our first example of a complex with this ligand architecture [3]. The tpy ligand is tridentate, and the pip_2NCN^- ligand is monodentate, bonded to Pt(II) through the phenyl ring. Electrochemical studies establish that this compound behaves as a nearly reversible two-electron reagent.

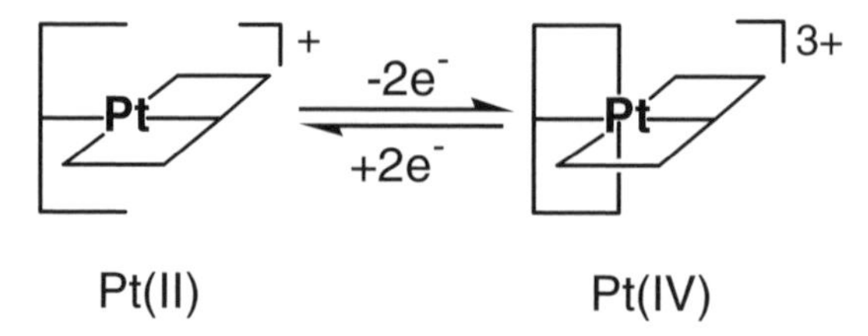

Figure 1. Schematic representation of a reversible two-electron platinum reagent.

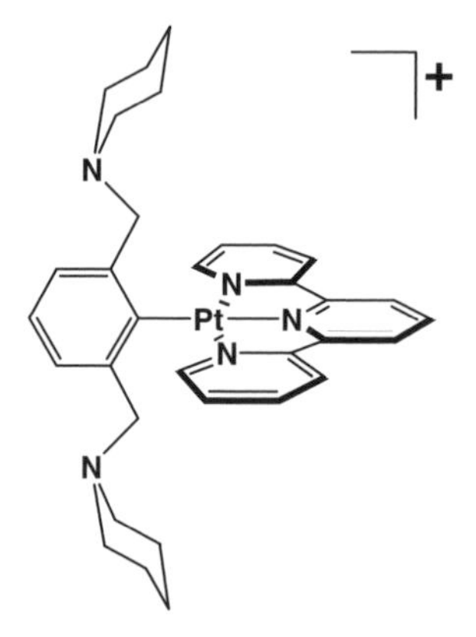

Figure 2. $[Pt(tpy)(pip_2NCN)]^+$ (**1**), a two-electron reagent.

EXPERIMENTAL

The compounds $[Pt(tpy)(pip_2NCN)](BF_4)$ (**1**(BF_4^-)), $[Pt(tpy)(pip_2NCNH_2)](PF_6)_3$ (**2**$(PF_6^-)_3$) and $[Pt(tpy)(dmph)](BF_4)$ (**3**(BF_4^-)) were prepared according to procedures that will be published elsewhere [3]. $Pt(pip_2NCN)Cl$ (**4**) was prepared as previously described [4]. All reagents were purchased from Acros or Aldrich. Acetonitrile was distilled over CaH_2. ^1H NMR spectra were recorded using a Bruker AC 250 MHz spectrometer. Deuterated solvents, $CDCl_3$ (0.03 % tetramethylsilane (TMS) (v/v)) and CD_3CN, were purchased from Cambridge Isotope Laboratories. Cyclic voltammetery was carried out using a standard three-electrode cell and a CV50w potentiostat from Bioanalytical Systems. Scans were collected in acetonitrile solution containing 0.1 M tetrabutylammonium hexafluorophosphate ($TBAPF_6$). All scans were recorded using a platinum wire auxiliary electrode, a Ag/AgCl reference electrode, and a 0.79 mm^2 gold working electrode. Between scans, the working electrode was polished with 0.05 μm alumina, rinsed with distilled water and wiped dry using a Kimwipe. Reported potentials are referenced against Ag/AgCl. Peak currents (i_p) and half-peak potentials ($E_{p/2}$) were estimated with respect to the extrapolated baseline current as described elsewhere [5]. The values of $(E_{pc}+E_{pa})/2$, which is an approximation of the formal potential for a redox couple, are referred to as E°'. Passed charge (Q) during the sweep was estimated by integrating the current from E°'+150 mV to E°'-150 mV. $TBAPF_6$ was recrystallized twice from methanol and dried in a vacuum oven prior to use. Calculated averages are reported as average value ± 2σ.

RESULTS AND DISCUSSION

The Pt(II) terpyridyl complexes **1–3** were isolated as salts and characterized by elemental analysis, ^{1}H NMR spectroscopy and mass spectrometry [3]. The tridentate coordination geometry of the tpy ligand was confirmed by ^{1}H NMR experiments. For each compound, the resonance for the aromatic proton of the tpy α carbon appears as a doublet with well resolved ^{195}Pt satellites ($J_{Pt\text{-}H}$=56, (**1**); 48, (**2**); 48 Hz (**3**)) resulting from three-bond coupling with the platinum center. Variable temperature studies of **1** indicate the presence of a single isomer in solution, establishing that the tpy ligand remains tridentate over the investigated temperature range (232–298 K) [3]. At room temperature, the resonance for the pip_2NCN^- benzylic protons appears as a singlet, whereas in the spectrum of the protonated complex (**2**), this resonance appears as a doublet (4.63 ppm, $J_{H\text{-}H}$=4 Hz). The absence of ^{195}Pt satellites for these resonances is indicative of the monodentate coordination geometry of the pip_2NCN^- ligand. When the pip_2NCN^- ligand is tridentate, as in $Pt(pip_2NCN)Cl$ (**4**), the benzylic resonance exhibits distinct ^{195}Pt satellites with $J_{Pt\text{-}H}$~50 Hz [3,4].

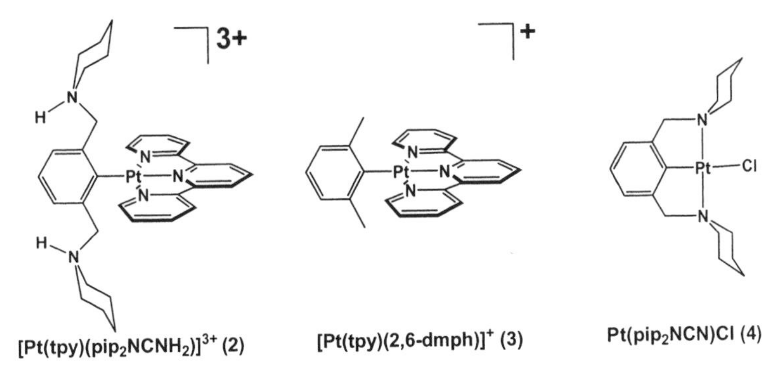

Figure 3. Related platinum(II) complexes with tpy or pip_2NCN^- ligands.

The structure of the chloride salt of **2** was confirmed in a preliminary X-ray crystallographic study (Figure 4). The terpyridine ligand is tridentate, and the pip_2NCN^- ligand is monodentate bonded through the phenyl ring. The phenyl group is nearly perpendicular (77.1(2)°) to the platinum coordination plane defined by the four atoms directly bonded to platinum. This orientation places the protonated amine groups above and below the coordination plane. Free rotation about the benzylic single bonds could, in principle, allow interaction between the piperdyl groups and the platinum center. However in crystals of the chloride salt of **2**, the amine groups are rotated away from the metal center, and there are no intramolecular N-H···Pt or agostic interactions. Weak N H···Pt interactions have been detected in the closely related $Pt(PCP)(Me_4NCNH_2)^{2+}$ complex (PCP=$(2,6\text{–}CH_2PPh_2)_2\text{–}C_6H_3^-$;$Me_4NCN^-$=$2,6\text{–}(CH_2NMe_2)_2\text{–}C_6H_3^-$) [6] as well as zwitterionic complexes such as *cis*-PtBr(C~N)(C~NH) (C~N=$Me_2N\text{-}CH(Me)\text{-}C_6H_4^-$) [7]. Weaker N-H···Pt interactions in **2** are consistent with reduced electron density on a Pt center bonded to the π-acceptor tpy ligand and only one anionic donor group.

A series of electrochemical experiments were undertaken to characterize the redox properties of **1**. The cyclic voltammagram of [Pt(tpy)(pip_2NCN)](BF_4) in acetonitrile solution (0.1 M $TBAPF_6$, 0.25 V/s) exhibits two reversible one—electron reduction waves at E°'= –0.98 V (i_{pc}/i_{pa}=0.80) and E°'= –1.50 V (i_{pc}/i_{pa}=0.52), with peak-to-peak separations (ΔE_p) of 65 and 61 mV, respectively (Figure 5). As expected for a Nernstian one-elec-

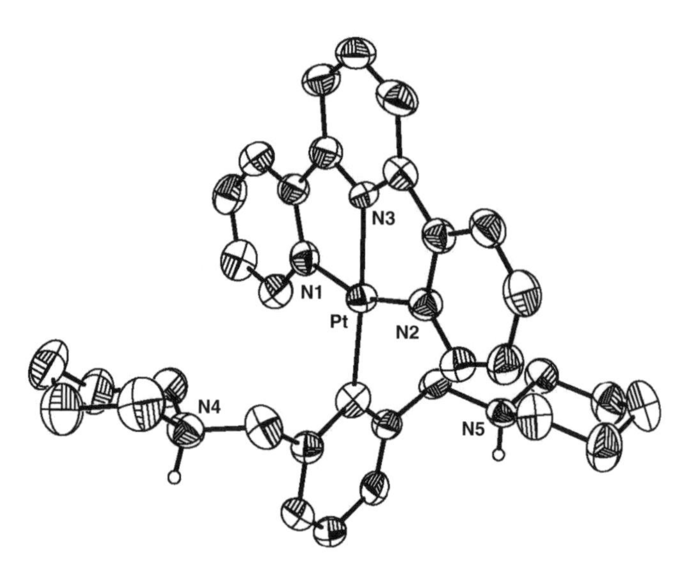

Figure 4. Ortep diagram of $Pt(tpy)(pip_2NCNH_2)^{3+}$. For clarity, all H atoms have been omitted with the exception of those bonded to N(amine) atoms.

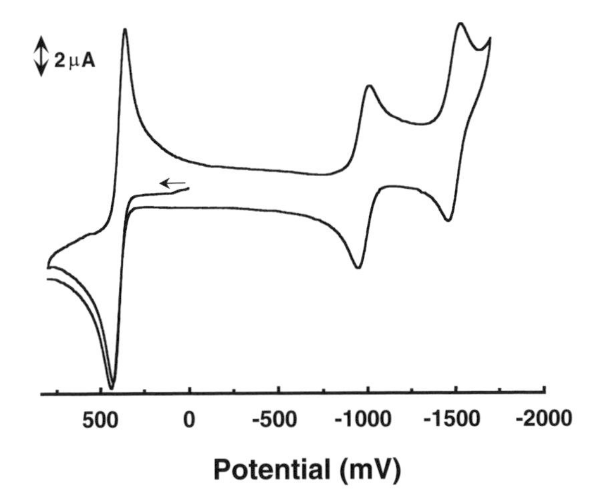

Figure 5. Cyclic voltammagram of $Pt(tpy)(pip_2NCN)][BF_4]$ (**1**(BF_4^-)) in acetonitrile (0.1 M $TBAPF_6$, 0.25 V/s).

tron process, the average value ΔE_p of the first reduction process (E°'=–0.98 V) is 59±6 mV for fourteen measurements with scan rates from 0.02 to 2.5 V/s. The complex also undergoes a nearly reversible two-electron oxidation process at E°'=0.40 V (i_{pc}/i_{pa}=1.08, ΔE_p=74 mV, 0.25 V/s) [8]. The peak currents of the forward and reverse sweeps are approximately twice those of the one-electron reduction process. At 0.25 V/s, the ratio of the peak anodic current of the oxidation process (i_{pa}=8.14 μA) to the peak cathodic current of the first reduction wave (i_{pc}=3.82 μA) is 2.1. Though somewhat less than 2.8 (=$2^{3/2}$), which is predicted for a Nernstian two-electron process, the ratio clearly exceeds the expected value (1.0) for a one-electron step. Similarly, the ratio of the charge passed during oxidation (Q_a) to the charge passed during the first reduction (Q_c) is 1.7.

Assignment of the redox processes observed for **1** can be inferred from comparison to the electrochemical properties of a series of related compounds, including those shown in Figure 3. Under identical conditions, neither free tpy, pip_2NCNBr nor $Pt(pip_2NCN)Cl$ (**4**) is reduced at potentials >–2.10 V, suggesting that the one-electron reduction processes for **1** are associated with the Pt(tpy) unit. Gray and coworkers [9] have observed that $Pt(tpy)Cl^+$ undergoes reversible one-electron reductions in DMF (0.1 M $TBAPF_6$) at E°'=–0.74 and E°'=–1.30 V, whereas $Zn(tpy)Cl_2$ undergoes reversible one-electron reduction at E°'=–1.36 V. The anodic shift of the ligand-centered couples in platinum(II) complexes is attributed to stabilization of the reduced tpy ligand as a result of coupling between the empty $6p_z$(Pt) and the π*(tpy) orbital [9]. As observed for **1**, $[Pt(tpy)(dmph)][BF_4]$

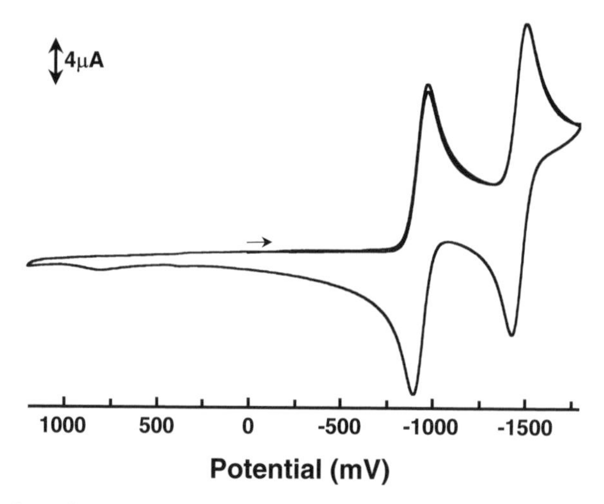

Figure 6. Cyclic voltammagram of **3** in acetonitrile (0.1 M $TBAPF_6$, 0.25 V/s).

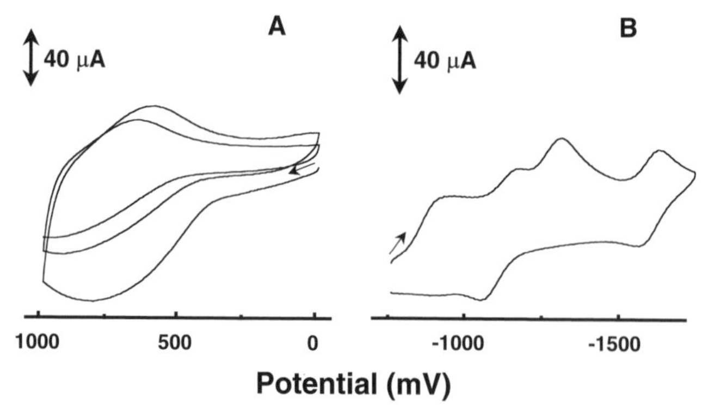

Figure 7. Cyclic voltammagrams of **2** in acetonitrile (0.1 M $TBAPF_6$, 0.25 V/s): (A) 2.5 cycles between 0 and 1000 mV, (B) one cycle between –400 and –1500 mV.

(**3**) in acetonitrile solution also undergoes two reversible one-electron reductions ($E°'$= 0.96 V, ΔE_p=60 mV, i_{pc}/i_{pa}=0.91; $E°'$= 1.49 V, ΔE_p=64 mV, i_{pc}/i_{pa}=0.90) (Figure 6). The small anodic shifts (10 and 20 mV, respectively) of these processes with respect to those observed for **1** are attributed to the influence of the electron donor properties of the methyl groups on the stability of the tpy anion. Conversely, the diammonium adduct (**2**) is reduced at more positive potentials (Figure 7b). A series of irreversible processes from –0.8 to –1.6 V are attributed to ligand-centered reductions, since the doubly protonated pip_2NCNBr adduct, $pip_2NCNBrH_2^{2+}$, is irreversibly reduced near –1.2 V, and the tpy anion is expected to react rapidly with protic acid.

The reversible two-electron oxidation wave observed for **1** is absent in cyclic voltammagrams of compounds **3** and **4**. Neither $Pt(tpy)(dmph)^+$ (**3**), pip_2NCNBr nor $pip_2NCNBrH_2^{2+}$ is oxidized at potentials <1.2 V [10]. $Pt(pip_2NCN)Cl$ (**4**) undergoes irreversible metal-centered oxidation near 0.8 V [4]. Taken together, these data indicate that both the pip_2NCN^- and tpy ligands play important roles in the unusal redox chemistry of **1**. The availability of the amine lone electron pairs is critical to facilitating reversible two-electron oxidation and stabilizing the resulting Pt(IV) center. For example, protonation of the piperdyl groups, as in compound **2**, results in oxidation near 0.4 V accompanied by electrode fouling (Figure 7a). This process most likely involves the platinum center as observed for **1**, but the protonation slows the overall kinetics and also causes increased surface adsorption.

In order to further characterize the electrochemical behavior of this system, cyclic voltammagrams were recorded for the first reduction process (–1.2 to –1.7 V) and the oxidation process (0.2 to 0.6 V) over a range of scan rates from 0.005 to 25.6 V/s. ΔE_p of the first reduction ($E°'$= 0.98 V) is essentially invariant (59±6 mV) for scan rates ranging from 0.02 to 2.5 V/s. Outside this range (0.01 0.015 and 5.1 25.6 V/s), the process is less reversible. As the scan rate increases from 5.1 to 25.6 V/s, ΔE_p increases from 66 to 88 mV, as expected from uncompensated resistance effects and the onset of slow electron-transfer kinetics relative to the scan rate. Below 0.01 V/s, the anodic and cathodic peak maxima were not resolved. Over the range of scan rates for which the electron transfer is under diffusion control (0.02 to 2.5 V/s), the average ratio of cathodic charge (Q_c) to the anodic charge (Q_a) is 3.8±1.7 for fourteen measurements. For faster scan rates, Q_c/Q_a lies near 3.0, as predicted for a chemically reversible process (*e.g.*, Q_c/Q_a=2.9, 2.5 V/s). With decreasing scan rate, the ratio increases (*e.g.*, Q_c/Q_a=5.7, 0.02 V/s), as expected for the onset decomposition of the reduced complex on these slow timescales. The variation and deviation of Q_c/Q_a with respect to the predicted ratio of 3.0 is partially a result of uncertainty in background current corrections. The cathodic peak current (i_{pc}) exhibits an approximately linear dependence on the square root of the scan rate (ν), as predicted by the Randles-Šev ik equation for Nernstian conditions [11,12]:

$$i_p = 2.69 \times 10^5 n^{3/2} A D^{1/2} C \nu^{1/2}, \quad (1)$$

where n=electron stoichiometry, A=electrode area, D=diffusion coefficient, and C=concentration (Figure 7). The slope of the linear fit over the range for which the reduction is electrochemically reversible (0.02 to 2.5 V/s) is 91.6 $\mu A(s/mV)^{1/2}$.

For the two-electron oxidation process, adsorption probably causes the slight anodic shift of E_{pa} and cathodic shift of E_{pc} during consecutive cycles.

As a consequence, ΔE_p increases by 2–4 mV for scan rates 1 V/s. For faster scan rates or using other electrode materials (glassy carbon or platinum) larger increases are observed during consecutive cycles, indicating further that the electron transfer is not purely outer-sphere and the process is complicated by the kinetics of adsorption and desorption.

At a gold electrode, ΔE_p for the first cycle increases continuously from 43 to 224 mV as the scan rate is increased from 0.01 to 20.5 V/s (Figure 8). In contrast to studies of $Ru(\eta^6{-}C_6Me_6)_2^{2+}$ by Geiger and coworkers [13], it was not possible to isolate individual one-electron steps comprising the overall two-electron reaction at fast scan rates, and the oxidation remains characterized by a single anodic and a single cathodic wave at even the fastest sweep rates. With decreasing scan rate, ΔE_p and E_p-$E_{p/2}$ (30 mV, 0.01 V/s) approach the two-electron Nernstian limits of 29.5 and 28.25 mV, respectively [14,15]. Thus, over the range of scan rates for which the peak-to-peak separation of the first reduction process is essentially invariant, ΔE_p for the oxidation process ranges from 43 to 150 mV. This behavior is consistent with a large structural reorganization resulting in slow reaction kinetics. At slow scan rates the process is less chemically reversible (*e.g.*, ΔE_p=43 mV, i_{pc}/i_{pa}=1.27, 0.01 V/s) indicating that the oxidized product is not long-lived at room temperature. This instability is not surprising, as we know of no known Pt(IV) tpy complexes in which the tpy ligand is tridentate [16,17]. At 273 K, the chemical reversibility improves, but the process is less electrochemically reversible as indicated by an increase in ΔE_p (*e.g.*, ΔE_p=91 mV, i_{pc}/i_{pa}=1.00, 0.01 V/s). While the lifetime of the oxidized product is improved at low temperature, the rate of electron transfer at the electrode is decreased, and the overall redox process is under increased kinetic control.

In order to verify the electron stoichiometry of the oxidation process, the anodic peak current (i_{pa}) is plotted against $\nu^{1/2}$ in Figure 9. Through the process clearly exhibits non Nernstian behavior as discussed earlier, the data are remarkably linear over the entire range of scan rates (0.005 to 25.6 V/s) as predicted by equation (1). The ratio of the slope of the best fit line (2.19 $\mu A(s/mV)^{1/2}$) to that obtained for the first reduction process provides an estimate of $(n_{ox}/n_{red})^{3/2}$. The resulting estimate of n_{ox}/n_{red} (=1.70) is consistent with the notion that oxidation of **1** involves transfer of two electrons per Pt center.

The accumulated data do not permit unambiguous identification of the mechanism of two-electron transfer. Nevertheless, the observed kinetic control of the oxidation reaction, as indicated by the dependence of ΔE_p on ν, and the structural rearrangement anticipated for the interconversion of Pt(II) and Pt(IV) suggest that the electron transfers are not concerted. Though the exact structure of the Pt(III) complex formed by removal of an electron from the d_{z2} level is uncertain, the resulting d^7 metal center in the presence of nucleophiles is expected to adopt a distorted five or six-coordinate structure [18–24]. For either case, the remaining unpaired electron may lie at

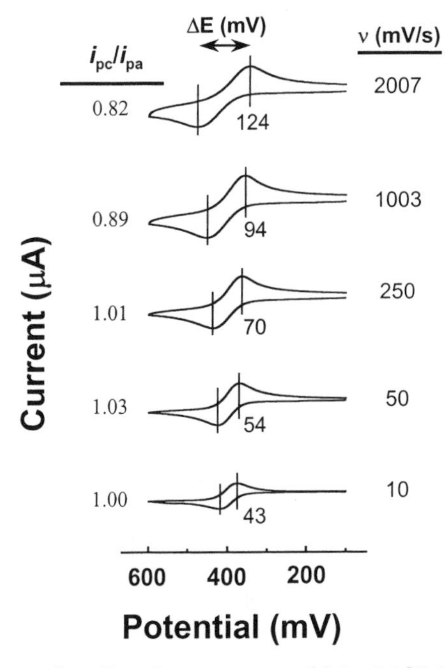

Figure 8. Dependence of cyclic voltammagram of 0.1 mM [Pt(tpy)(pip$_2$NCN)][BF$_4$] (**2**) in acetonitrile on scan rate (0.1 mM TBAPF$_6$). Scans are scaled by factors of 1 (2.0), 1.33 (1.0), 2 (0.25), 3 (0.05), and 4 (0.01 V/s), respectively.

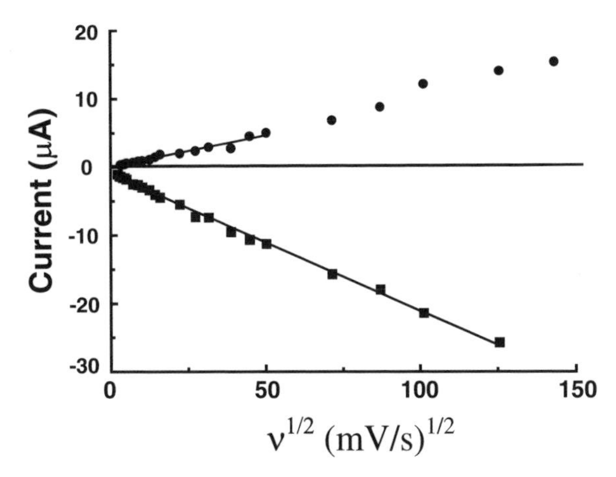

Figure 9. Dependence of anodic peak current (i_{pa}) for 0.4 V oxidation process (·) and cathodic peak current (i_{pc}) for 0.98 V reduction process (·) on the square root of scan rate ($\nu^{1/2}$) for [Pt(tpy)(pip$_2$NCN)][BF$_4$] (**1**(BF$_4^-$)), in acetonitrile (0.1 mM TBAPF$_6$). Scans recorded from 0.2 to 0.6 V and from 1.2 to 1.7 V. Lines represent linear fits of all oxidation data and reversible reduction data (0.02–2.5 V/s).

higher energy than the d_{z^2} level in the Pt(II) complex, allowing for oxidation at low potential and formation of a transiently stable six-coordinate Pt(IV) product. Thus, ligation accompanying the first electron transfer effectively drives the second charge-transfer step. Accordingly, the observation that ΔE_p and E_{pa}-$E_{pa/2}$ exceed the Nernstian two-electron values at slow scan rates is consistent with stepwise electron-transfer in which the second electron-transfer step is slightly more favorable than the first (*i.e.*, $E_2^{\circ\prime}>E_1^{\circ\prime}$) [14]. It also is noteworthy that the anodic shift of E_{pa} with increasing scan rate is smaller than the cathodic shift of E_{pc}. As a consequence, $E^{\circ\prime}$ shifts from 0.40 V at 0.005 V/s to 0.44 V at 20 V/s. This behavior is consistent with formation of an organized structure, such as a five or six-coordinate complex, prior to or during anodic electron transfer. The conversion of Pt(IV) to Pt(II) and axial bond cleavage during the reverse (cathodic) process is likely to require less pre-organization, making E_{pc} less sensitive to scan rate.

The electrochemistry of **1** is distinctly different from that of other platinum(II) complexes, the majority of which undergo electrochemically and chemically irreversible oxidation because of accompanying changes in coordination sphere. However, in the presence of high concentrations of coordinating ligands, the cyclic voltammagram can exhibit a two-electron wave with large ΔE_p, as observed for **1** [25,26]. For example, square planar $Pt([14]aneS_4)^{2+}$ ($[14]aneS_4$=1,4,7,11–tetrathiacyclotetradecane) in a 0.1 M LiCl 4:1 CH_3CN:H_2O solution undergoes two-electron oxidation at 0.8 V (ΔE_p=127 mV, 0.01 V/s) [25]. In contrast, cyclic voltammagrams of complexes with one or two pendant groups available to bind to the oxidized metal center, such as $Pt([9]aneS_3)_2^{2+}$ ($[9]aneS_3$=1,4,7–trithiacyclononane), exhibit one-electron waves corresponding to generation of a five- or six-coordinate Pt(III) species [23,27–30]. In some cases, it is possible to resolve the second one-electron Pt(III)/Pt(IV) step at higher potentials. Similarly, steric protection of the axial sites of the metal complex allows reversible formation of a four-coordinate Pt(III) complex by shifting the Pt(III)/Pt(IV) couple to higher potentials [19,20,31]. The contrasting facile Pt(II)/Pt(IV) interconversion observed for **1** may be in part related to the relative instability of a Pt(III) center bonded to ligands that favor small bite angles and meridional coordination geometries. The role of the pip_2NCN^- ligand in facilitating electron transfer is further suggested by the comparison with the electrochemistry of $W_2(CO)_8(\mu\text{-SBz})_2^{2-}$ reported by Schultz [32,33]. The dimer undergoes nearly reversible oxidation at –0.87 V vs. Ag/0.01 M $AgNO_3$ (ΔE_p=42 mV, 0.2 V/s), accompanied by formation of a metal-metal bond and modest structural rearrangement (~0.1 Å decrease in W-W distance and ~25° change in W S W and S W S bond angles) [32,33]. Though the variation of ΔE_p with scan rate (104 mV, 20V/s) is somewhat less pronounced than observed for **1**, values of ΔE_p and E_{pa} $E_{p/2}$ (37 and 32 mV, respectively, 0.005 V/s) are similar to those observed for **1** at slow scan rates. It is conceivable that weakly favorable interactions between the pip_2NCN^- amine groups and the Pt(II) center in **1** effectively preorganize the complex for electron transfer.

ACKNOWLEDGEMENTS

H. J. wishes to thank Professor William B. Connick for his support and guidance, and Professor Harry B. Mark, Dr. Jeanette A. Krause Bauer, Dr. E. Brooks and Dr. M. J. Goldcamp for helpful discussions and expert technical assistance. Diffraction data were collected by Dr. Krause Bauer through the Ohio Crystallographic Consortium, funded by the Ohio Board of Regents 1995 Investment Fund (CAP-075) and located at the University of Toledo, Instrumentation Center in A&S, Toledo, OH 43606. H.J. is grateful to the University of Cincinnati University Research Council for summer research fellowships and a Distinguished Dissertation Fellowship, the University of Cincinnati Department of Chemistry for the Stecker Fellowship, and the Link Foundation for an Energy Fellowship.

REFERENCES

[1] A. F. Heyduk; D. G. Nocera "Hydrogen Produced From Hydrohalic Acid Solutions by a Two-Electron Mixed-Valence Photocatalyst" *Science 293(5535)*, 1639–1641 (2001).

[2] G. Lappin *Redox Mechanisms in Inorganic Chemistry*; Ellis Horwood: New York (1994).

[3] H. Jude; J. A. Krause Bauer; W. B. Connick, *manuscript in preparation.*

[4] H. Jude; J. A. Krause Bauer; W. B. Connick "Synthesis, Structures, and Emissive Properties of Platinum(II) Complexes with a Cyclometallating Aryldiamine Ligand" *Inorg. Chem 41*, 2275–2281 (2002).

[5] P. T. Kissinger; W. R. Heineman "Cyclic Voltammetry" *J. Chem. Edu. 60*, 702–706 (1983).

[6] M. Albrecht; P. Dani; M. Lutz; A.L. Spek; G. van Koten "Transcyclometalation Processes With Late Transition Metals: C-Aryl-H Bond Activation Via Noncovalent C-H . . . Interactions" *J. Am. Chem. Soc. 122*, 11822–11833 (2000).

[7] I. C. M. Wehman-Ooyevaar; D. M. Grove; H. Kooijman; P. van der Sluis; A. L. Spek; G. van Koten "A Hydrogen Atom in an Organoplatinum-Amine System 1. Synthesis and Spectroscopic and Crystallographic Characterization of Novel Zwitterionic Complexes With a Pt(II)···H-N+ Unit" *J. Am. Chem. Soc. 114*, 9916–9924 (1992).

[8] Oxidation of $[Pt(tpy)(pip_2NCN)](BF_4)$ is less reversible using platinum (i_{pc}/i_{pa}=0.93, ΔE_p=160 mV, 0.25 V/s) or glassy carbon electrodes (i_{pc}/i_{pa}=0.75, ΔE_p=300 mV, 0.25 V/s), or CH_2Cl_2 solvent (i_{pc}/i_{pa}=0.72, ΔE_p=320 mV, 0.25 V/s, gold electrode, 0.1 mM TBAPF6).

[9] J. A. Bailey; M. G. Hill; R. E. Marsh; V. M. Miskowski; W. P. Schaefer; H. B.

Gray "Electronic Spectroscopy of Chloro(terpyridine)platinum(II)" *Inorg. Chem. 34,* 4591–4599 (1995).

[10] An adsorption wave occurs near 0.8 V. The current is insensitive to concentration, and this feature is absent when using a glassy carbon electrode.

[11] J. E. B. Randles "A Cathode Ray Polargraph. Part II—The Current-Voltage Curves" *Trans. Faraday Soc. 44,* 327–338 (1948).

[12] A. Ševčik "Oscillographic Polarography With Periodical Triangular Voltage" *Coll. Czech. Chem. Commun. 13,* 349–377 (1948).

[13] D. T. Pierce; W. E. Geiger "Splitting a Two-Electron Cyclic Voltammetric Wave Into Its One-Electron Components: The $(\eta\text{-}C_6Me_6)_2Ru^{2+/+/0}$ Couples" *J. Am. Chem. Soc. 111,* 7636–7638 (1989).

[14] R. L. Myers; I. Shain "Determination of $E_2°\text{-}E_1°$ for Overlapping Waves in Stationary Electrode Polarography" *Anal. Chem. 41,* 980 (1969).

[15] A. J. Bard; L. R. Faulkner *Electrochemical Methods*; Wiley: New York (1980)

[16] A. Gelling; K. G. Orrell; A. G. Osborne; V. Sik "The Energetics and Mechanism of Fluxionality of 2,2':6',2''-Terpyridine Derivatives When Acting as Bidentate Ligands in Transition-Metal Complexes. A Detailed Dynamic NMR Study." *J. Chem. Soc. Dalton Trans.* 937–935 (1998).

[17] A. Gelling; M. D. Olsen; K. G. Orrell; A. G. Osborne; V. Sik "Synthesis and Dynamic NMR Studies of Fluxionality in Rhenium(I), Platinum(II), and Platinum(IV) complexes of 'Back-to-Back' 2,2':6',2''-Terpyridine Ligands." *J. Chem. Soc., Dalton Trans.* 3479–3488 (1998).

[18] R. Uson; J. Fornies; M. Tomas; B. Menjon; K. Suenkel; R. Bau "The First Mononuclear Platinum(III) Complex. Molecular Structures of $(NBu_4)[Pt^{III}(C_6Cl_5)_4]$ and of its Parent Compound $(NBu_4)[Pt^{III}(C_6Cl_5)_4]\cdot 2CH_2Cl_2$" *J. Chem. Soc., Chem. Commun. 12,* 751–752 (1984).

[19] R. Uson; J. Fornies; M. Tomas; B. Menjon; R. Bau; K. Enkel; E. Kuwabara "Synthesis and reactivity of $[NBu_4]^+[Pt^{III}(C_6Cl_5)_4]^-$: Molecular Structures of $[NBu_4]_2^+[Pt(C_6Cl_5)_4]^{2-}\cdot 2CH_2Cl_2$, $[NBu_4]+[Pt^{III}(C_6Cl_5)_4]^-$ and $[NBu_4]^+[Pt(C_6Cl_5)_4(NO)]^-$" *Organometallics 5,* 1576–1581 (1986).

[20] J. Fornies; B. Menjon; R. M. Sanz-Carrillo; M. Tomas; N. G. Connelly; J. G. Crossley; A. G. Orpen "Synthesis and Structural Characterization of the First Isolated Homoleptic Organoplatinum(IV) Compound: $[Pt(C_6Cl_5)_4]$" *J. Am. Chem. Soc. 117,* 4295–4304 (1995).

[21] A. Goursot; H. Chermette; W. L. Waltz; J. Lilie "Theoretical Study by the Xα Method of Platinum(III) Complex Ions Containing Aquo and Chloro Ligands" *Inorg. Chem. 28,* 2241–2247 (1989).

[22] W. L. Waltz; J. Lilie; A. Goursot; H. Chermette "Photolytic and Radiolytic Study of Platinum(III) Complex Ions Containing Aquo and Chloro Ligands" *Inorg. Chem. 28,* 2247–2256 (1989).

[23] A. M. Bond; R. Colton; D. A. Fiedler; J. E. Kevekordes; V. Tedesco; T. F. Mann "Voltammetric and Spectroscopic Studies Related to Platinum(II) and Platinum(IV) Dithiocarbamate Redox Chemistry: Electrochemical, ESR, and Electrospray Mass Spectrometric Identification of a Moderately Stable Platinum(III) Cation: $[Pt(S_2CNEt_2)(\eta^3\text{–}P_2P')]^{2+}$ (P_2P' = $Ph_2P(CH_2)_2P(Ph)(CH_2)_2PPh_2$)" *Inorg. Chem. 33,* 5761–5766 (1994).

[24] M. Geoffroy; G. Bernardinelli; P. Castan; H. Chermette; D. Deguenon; S. Nour;

J. Weber; M. Wermeille "The Oxidation Reaction in X-irradiated Bis(ethylenediamine)platinum(II) Bis(hydrogen squarate): A Single-Crystal EPR Study of a Platinum(III) Complex" *Inorg. Chem. 31,* 5056–5060 (1992).

[25] M. A. Watzky; D. Waknine; M. J. Heeg; J. F. Endicott; L. A. Ochrymowyzc "Tetradentate macrocyclic complexes of platinum. Evaluation of the stereochemical alterations of redox behavior and the X-ray crystal structure of (1,4,7,10–tetrathiacyclododecane)platinum(II) chloride." *Inorg. Chem. 32,* 4882–4888 (1993).

[26] A. T. Hubbard; F. C. Anson "Study of electrochemistry of chloride and bromide complexes of platinum(II) and -(IV) by thin-layer electrochemistry." *Anal. Chem. 38,* 1887–1893 (1966).

[27] A. J. Blake; R. O. Gould; A. J. Holder; T. I. Hyde; A. J. Lavery; M. O. Odulate; M. Scrhöder "Stabilization of Trivalent Platinum by Structurally Accommodating Thiamacrocylces" *J. Chem. Soc. Dalton Trans.* 118–20 (1987).

[28] A. J. Blake; R. D. Crofts; M. Schröder "Synthesis, structure and electrochemistry of $[Pt([10]anes3)_2][PF_6]2$ ([10]anes3 = 1,4,7–trithiacyclodecane)" *J. Chem. Soc. Dalton Trans.* 2259–60 (1993).

[29] G. J. Grant; N. J. Spangler; W. N. Setzer; D. G. VanDerveer; L. F. Mehne "Synthesis and complexation studies of mesocyclic and macrocyclic polythioethers. XIV. Crown thioether complexes of palladium(II) and platinum(II)." *Inorg. Chim. Acta 246,* 31–40 (1996).

[30] R. I. Haines; D. R. Hutchings; T. M. McCormack "Platinum carboxylato-pendant-arm macrocycles: structure, redox properties and anti-cancer potential." *J. Inorg. Biochem. 85,* 1–7 (2001).

[31] A. Klein; W. Kaim "Axial Shielding Of 5d(8) and 5d(7) Metal Centers In Dimesitylplatinum Complexes With Unsaturated Chelate Ligands—Spectroscopic and Spectroelectrochemical Studies Of 4 Different Oxidation-States" *Organometallics 14,* 1176–1186 (1995).

[32] D. A. Smith; B. Zhuang; W. E. Newton; J. W. McDonald; F. A. Schultz "Two-Electron Transfer Accompanied by Metal-Metal Bond Formation. Synthesis and Electrochemistry of Dinuclear Molybdenum and Tungsten Carbonyl Thiolates" *Inorg. Chem. 26,* 2524–2531 (1987).

[33] J. B. Fernandes; L. Q. Zhang; F. A. Schultz "Correlation of Heterogeneous Electron Transfer Rate with Structural Change and Environmental Factors in the Two-Electron Oxidation of $W_2(CO)_8(\mu\text{-}SBz)_22–$" *J. Electranal. Chem. 297,* 145–161 (1991).

Hydrogen Adsorption Studies on as Produced Single Walled Carbon Nanotubes: Implications on Energy Application

Saikat Talapatra

Department of Physics

Southern Illinois University at Carbondale

Carbondale, IL 62901-4401

Research Advisor: Prof. A. D. Migone

ABSTRACT

We present adsorption isotherm study of hydrogen on as produced Single walled carbon nanotube bundles (SWNTs). Measurements were performed at seven different temperatures between 25 K–60 K to determine the isosteric heat of adsorption and the specific surface area (SSA) of the SWNT samples. The values of isosteric heat at different coverages were measured and the corresponding adsorbate-substrate binding energy value was determined. The SSA and the binding energy value obtained using hydrogen as an adsorbate is compared with the same quantities obtained for other gas species (Xenon, Argon etc.). In light of the results obtained the feasibility of as produced SWNT's for hydrogen storage for fuel cell application is discussed.

INTRODUCTION AND BACKGROUND

The ease of production and the possibility of developing hydrogen as an environmentally friendly, convenient fuel for transportation have led to the study of hydrogen storage systems as one of the most focused field of scientific research and development in the past few decades [1]. The main reason is the pace of innovation in fuel cells, which are, in essence, batteries that use hydrogen to produce electrical energy efficiently. This is an important component of the 21st century "hydrogen driven" economy leading to a greener, clean environment. There is one important question, however, that needs to be answered satisfactorily for this technology to come to fruition: how exactly will hydrogen be stored? One way is the physical storage of hydrogen, as a compressed gas or in liquefied form. Another way is the chemical storage of hydrogen, in such fuels as methanol. Both approaches would require expensive investment in fuel infrastructure. Apart from these two methods Gas-on-solid adsorption is also a popular method for automotive hydrogen storage system.

With the discovery of single walled carbon nanotubes [2] (SWNTs) and the initial reports of high reversible adsorption of molecular hydrogen [3] in them, have made these materials a very promising medium for hydrogen storage for fuel cell applications. A SWNT can be viewed as a single graphene sheet rolled over itself and closed seamlessly, with its ends capped. SWNT's assemble into bundles. Theoretical studies have identified three different types of adsorption sites in close-ended SWNT bundles: i.- the cylindrical outer surfaces of the individual nanotubes that lie at the external surface of the bundles; ii.- the region where two of these external tubes come close together, i.e. the outer "grooves" on the surface of the bundle; and, iii.- the space in between the individual nanotubes at the interior of the bundle, i.e. the interstitial channels or IC's [4]. A schematic of a cross section of a nanotube bundle indicating these three different groups of sites is shown in Figure.1. As produced bundles of SWNT's have at least two sites (the "grooves", and, the outer surface of individual tubes) that are available for adsorption for all gases. The IC's, in theoretical studies have been identified as potential sites capable of adsorption within the bundle. In practice the adsorption in the IC's in as produced SWNT's is severely restricted by the size of the adsorbates.

Hydrogen adsorbed on carbon nanotubes have been suggested as the basis for a new gas storage technology that could lead, if it comes to fruition, to a complete revolution in the automotive industry by enabling the economically competitive production of hydrogen powered vehicles [5]. The U.S. Department of Energy (DOE) hydrogen plan has declared as a commercially significant bench mark for the amount of reversible hydrogen adsorption a ratio of stored hydrogen weight of 6.5 wt% hydrogen

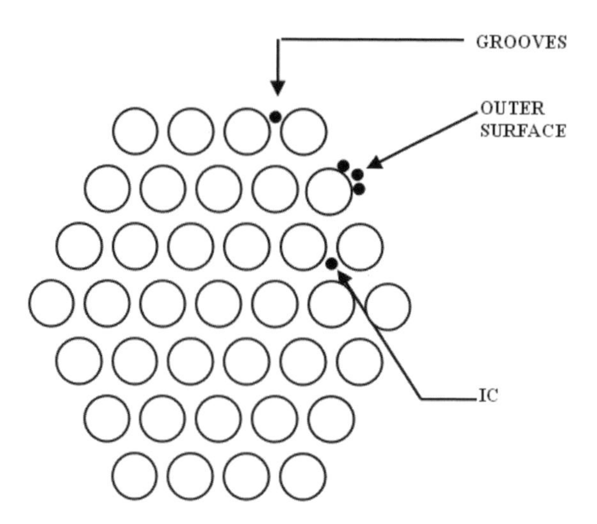

Figure 1. Schematic representation of available adsorption sites in as produced single walled carbon nanotubes.

and, and a volumetric density of 62 kg hydrogen/m^3. Some encouraging preliminary results suggested that it may be possible to reach the goals of DOE hydrogen plan.

Because of the potentially very large practical payoff that would result from the development of an economically competitive, effective means of achieving H_2 storage at room temperature (for fuel storage in automotive) applications, much of the work has focused up to now on measurements a room temperature, or higher; and generally it is conducted at higher pressures. Recent results, obtained at high pressures (about 100 atmospheres), on the H_2 storage capacity of some of these new carbons are remarkable in that they defy, both, current expectations for the storable amounts, as well as theoretical attempts at understanding the results based solely on physisorption phenomena [6–8]. Therefore, establishing the adsorptive capacity of hydrogen on these new forms of carbon is presently a matter of considerable controversy and much activity. It is believed that considerable measure of clarity on this subject will be gained from the experimental determination of the binding of H_2, at low pressures and low temperatures.

Keeping this fact in mind adsorption isotherm measurements of hydrogen were performed on SWNT's at low temperatures for determining effective area available for adsorption and the binding energy of hydrogen on these samples.

ADSORPTION ISOTHERM MEASUREMENT (THEORETICAL BASIS)

An adsorption isotherm is a determination of the amount of the gas atoms adsorbed on the substrate, as a function of the equilibrium pressure of the gas P, at a constant temperature. By measuring the initial pressure p_i and the final pressure p_f in the gas handling system and knowing the volume of the system V, we can calculate the amount of gas admitted into the sample cell A_{in} as:

$$A_{in} = (p_I - p_f)V \cdot \quad (1)$$

By adding the amounts of gas admitted in together, we know the total amount of gas in the cell A_{sum}:

$$A_{sum} = \quad A_{in} \cdot \quad (2)$$

Then by subtracting the amount of gas atoms present in the vapor phase inside the experimental cell, we can get the amount of atoms adsorbed on the surface, N, as:

$$N = A_{sum} - p_f V_{eff}, \quad (3)$$

where V_{eff} is the effective volume of the cell (V_{eff} is determined by the helium expansion into the cell and applying the ideal gas law as if it were at room temperature).

From the above equations we can see that what we measure to get an adsorption isotherm is mainly the 3–D pressure of the gas. So, how is that by measuring 3–D quantities we can get information on 2–D or 1–D quantities? To get an insight into the connection between the 3–D vapor phase and lower dimension film of the gas we must look into the equilibrium condition between the vapor and the film.

The differential Gibbs function for the film (let us assume that we are dealing with 2–D film) and the 3–D vapor are respectively,

$$dG_{film} = -S_{film}\,dT + A d\phi + \mu_{film}\,dN_{film} \cdot \quad (4)$$

(Here A is the area of the film and f is the 2–D pressure, also known as the spreading pressure).

$$dG_{vapor} = -S_{vapor}\,dT + VdP + \mu_{vapor}\,dN_{fvapor} \cdot \quad (5)$$

We know that [9]

$$G_{film} = \mu_{film} N_{film} , \tag{6}$$

$$G_{vapor} = \mu_{vapor} N_{vapor} \tag{7}$$

this implies;

$$dG_{film} = \mu_{film} dN_{film} + N_{film} d\mu_{film} , \tag{8}$$

$$dG_{vapor} = \mu_{vapor} dN_{vapor} + N_{vapor} d\mu_{vapor} . \tag{9}$$

Setting (2.1.4) = (2.1.8) and (2.1.5) = (2.1.9) we have;

$$N_{film} d\mu_{film} = -S_{film} dT + Ad\phi , \tag{10}$$

$$N_{vapor} d\mu_{vapor} = -S_{vapor} dT + VdP . \tag{11}$$

In equilibrium: $d\mu_{film} = d\mu_{vapor}$,
or,

$$- s_{film} dT + a_{film} df = -s_{vapor} dT + v_{vapor} dP \tag{12}$$

In an isotherm $dT = 0$ therefore the equilibrium condition becomes;

$$ad\phi = vdP , \tag{13}$$

which essentially means the chemical potential of the film at equilibrium is equal to the chemical potential of the vapor 3–D pressure. This last equation connects the 3–D quantities with 2–D quantities or 1–D quantities.

AUTOMATED ISOTHERM MEASURING SETUP

The isotherm measuring set up was fully computer controlled. Figure 2 shows the schematic of the apparatus. The apparatus used includes a gas handling system, various pressure gauges with different ranges (1 Torr, 10 Torr, 100 Torr and 1000 Torr) and a temperature controlled cell arranged as shown in the figure 3. Three electro-pneumatic valves (valve #1, valve #2, and valve #3) controlled by IBM-PC compatible computer were used in the gas handling system, all other valves are manual. The electro-pneumatic valves can be operated either by computer, or manually, allowing the automation of the adsorption measurement.

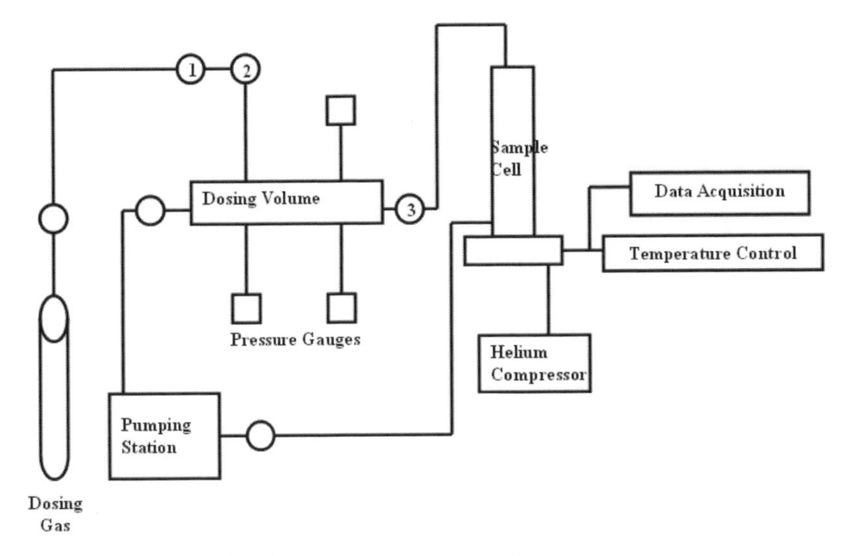

Figure 2. Automated Adsorption Isotherm Set-Up.

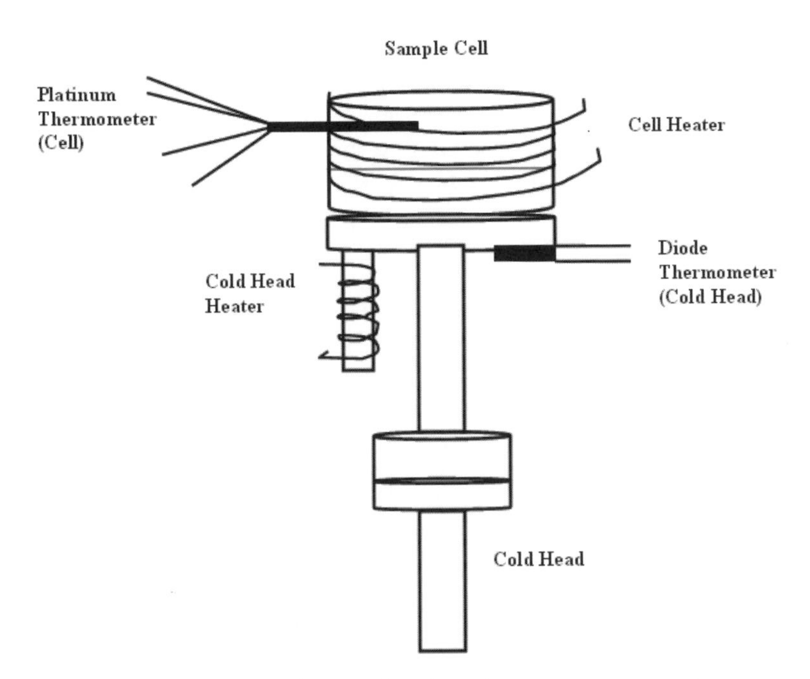

Figure 3. Arrangement of the sample cell and temperature control.

The sample is placed inside the cell. The reference volume is measured before being connected to the gas handling system, and it is used to determine the volume of the system by using gas expansion into the system and applying ideal gas law. The temperature of the cell is determined by the cold head, cold head heater and the cell heater. A schematic diagram of the temperature measurement and the heating system is shown in Figure 3. The cooling power of the cold head is fixed, but the powers of the two heaters are controlled by the two independent temperature controllers. With this arrangement the cell can have a temperature stability of (~±0.005°K), and it can be controlled over a wide range of temperatures (from 19° K to room temperature). The temperature of the cold head and the cell are displayed in the temperature controller and the cell temperature is recorded in the computer by the interface.

The pressure of the gas in the system is converted into a voltage signal by the capacitance gauges. It is measured by a voltmeter, and, is recorded by the computer. The equilibrium condition for the pressure is established by having the differences in pressure recorded at a given particular time interval being smaller than a value preset at the beginning of the run (typically, when pressure is lower than 0.5 Torr, it is 0.0005 Torr, and when the pressure is higher than 0.5Torr, it is 0.1% of the pressure over a time interval of 90 seconds after having had the system "sleeping" for a preset interval (of at least 1,700–20,000 secs). In our case typical waiting or "sleeping time" were on the order of 10,000 secs.

It is through the measurements and control system that the computer records the cell temperature and the pressure of the gas inside the gas handling system. This same measurement and the control system operate the electro-pneumatic valves. The measurement and the control system used is Keithley KM488 DD, the data acquisition and the control software is DAS 8. The program used to run the adsorption is named "iso1" and was developed in-house, the program runs under Microsoft GWBASIC.

Assuming that an initial amount of gas has been admitted to the cell, the program includes two parts: prompting for parameters which are needed to control the run, and repeating cycle of run, which represents one data point in the adsorption isotherm. The run stops when the total amount of gas admitted to cell reaches a preset value.

At the beginning of each run, a shot of gas between valve#1 and valve#2 is admitted to the gas handling system through opening and closing of these two valves. Then the equilibrium pressure of the gas handling system is measured as initial pressure p_I and the amount of gas inside the gas handling system is calculated. If the amount of gas inside the gas handling system is smaller than the preset value, another shot of gas is admitted and the new p_I is measured. Once the preset shot size is reached p_I is recorded and the valve#3 (to the sample cell) is opened. After valve#3 is opened and after waiting for a preset time for attaining the equilibrium, the final equilibrium pressure p_f in the gas handling system is measured and the valve#3 is closed. Then the amount of gas admitted to the cell and the amount of gas adsorbed are calculated from the pressure differences. All the measured and calculated values are stored in the data file. The process is then repeated. Once the amount dosed into the cell reaches the preset value, the run is complete.

SUBSTRATES AND ADSORBATES

The sample used in our experiments was as produced arc discharge SWNT bundles prepared by C.Journet at Prof. Bernier's lab. The reported purity of the sample is of the order of 80%. The nanotubes have a typical diameter of 1.38nm, forming a triangular lattice of 1.7nm, which gives an estimated IC diameter of .26nm, excluding the electron cloud. A detailed discussion of the synthesis of the sample is given elsewhere [10]. The nanotubes were held in a container open to atmosphere after production. The sample was placed in a copper cell and evacuated to pressures lower than 1×10^{-6} Torr for a period of at least 12 hours, at room temperature, prior to the performance of the adsorption measurements. .29 grams of samples were used for all the measurements reported in the present study.

The adsorbates used in this study were 99.999% pure gases (xenon, argon and hydrogen and neon) produced by Matheson.

RESULTS AND DISCUSSION

Most of the experimental studies reported so far were mainly focused on the storage capacity of these materials and were performed at or near room temperatures and at above atmospheric pressures. Very few experimental studies have focused on the low temperature adsorption behavior, the sites available for adsorption, and the nature of adsorption of molecular hydrogen on SWNT's.

Inelastic neutron scattering performed by C.M.Brown et.al [11] on as produced SWNT's concluded that main binding sites for physisorbed hydrogen is the exterior bundle surface. This study also provided an estimate for the binding energy value of the adsorbed hydrogen. A recent volumetric adsorption isotherm measurement study [12] of hydrogen on as produced SWNT with in temperatures ranging from 32 K–90 K, showed for the first time the presence of two sharp features in the isotherm data suggesting at least two different sites of adsorption for hydrogen, present at low temperatures and pressures, on SWNT bundles which in that study were identified respectively as grooves/interstitials (high energy) and exterior (low energy). The isosteric heats of adsorption associated with each site was calculated and was used to provide an explanation regarding the location of the adsorbed hydrogen. A very recent [13] Raman spectroscopic investigation of hydrogen adsorbed on SWNT at 85 K and pressures up to 8 atm was also analyzed in terms of two different binding sites for hydrogen. In this study by considering the shifts in the Raman frequencies, the possible sites of hydrogen physisorbed on SWNT's are discussed. The shifts in the experimental Raman data were compared with the shifts of the adsorbate stretching modes estimated in two geometries present in SWNT bundles (interstitial channels and the outer bundle surface) from a model potential. All these three experimental findings are consistent in their conclusion regarding one group of site for hydrogen adsorption viz. on the outer surface of the SWNT's and the nature of the adsorption. The nature of the second group of binding sites for hydrogen on SWNT's ("groove/interstitial" in the volumetric study and "interstitial channels" in the Raman scattering investigation) is an issue, discussed in this section.

We have calculated the specific surface area (SSA) offered for hydrogen on the sample under investigation. SSA measurements on the same sample were also done using other gases (Xenon and Argon) that are bigger than Hydrogen. These gases especially Xe, are fully expected to be larger than the most reasonable estimates for the IC's, as well. The results of the SSA obtained with each of the gases are quite similar; this suggests very strongly that for as produced SWNT's hydrogen adsorption does not occur in the IC's.

Figure 4 shows adsorption isotherms measurements performed at temperatures ranging between 25 K-65 K. Amount adsorbed in cc-Torr is plotted on X-axis as function of logarithm of pressure in Torr (Y-axis). The lower temperatures isotherms begin at higher coverage because the pressure values at which low coverage adsorption begins are sufficiently low and the gauges used in our setup are unable to determine these pressure values accurately.

The series of adsorption isotherms shown in Figure 4 can be used to determine the isosteric heat of adsorption at different coverages. The isosteric heat of adsorption is the amount of heat released when an atom or molecule is adsorbed on a substrate. It is defined as: $q_{st} = kT^2(\partial \ln P/\partial T)_N$ where T is temperature in Kelvin, P is the equilibrium pressure and N is the amount of adsorbed atoms. We plotted LnP vs. 1/T for different, fixed, values of the coverage, N, for data taken from the isotherms in figure 4. The resulting curves are essentially straight lines. The slopes of these straight lines, multiplied by an appropriate factor, determines the isosteric heat of adsorption at the particular value of the coverage selected.

Figure 5 shows an adsorption isotherm of hydrogen on SWNT's at 40.2 K. Amount adsorbed in cc-Torr (X-axis) as function of pressure in Torr (Y-axis).

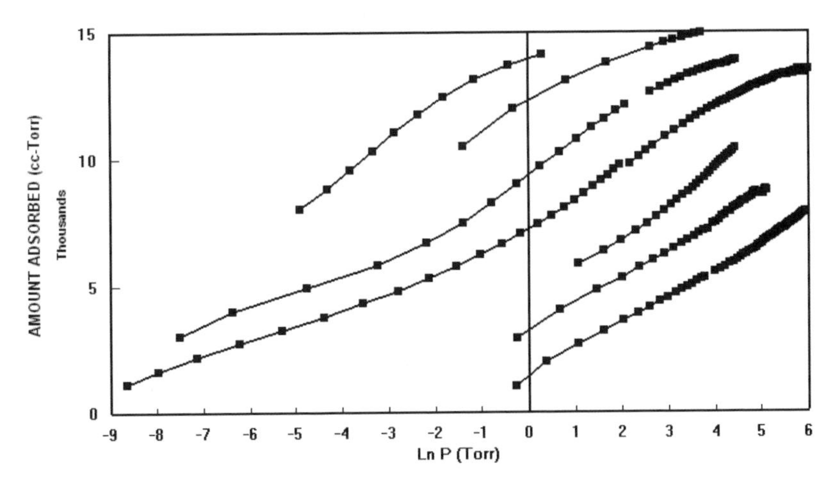

Figure 4. Hydrogen adsorption on SWNT. Temperatures from left are 25.36 K, 29.2 K, 35.2 K, 40.2 K, 50.18 K, 55 K, and 65.14 K.

This isotherm shows a monolayer completion of hydrogen on the SWNT sample. The monolayer completion (which occurs at about 12,250 cc-Torr) was determined by extrapolating the straight line portion of the isotherm. This monolayer value and the isotherms shown in figure 4 were used to calculate the isosteric heats of adsorption at fraction monolayer coverage. We

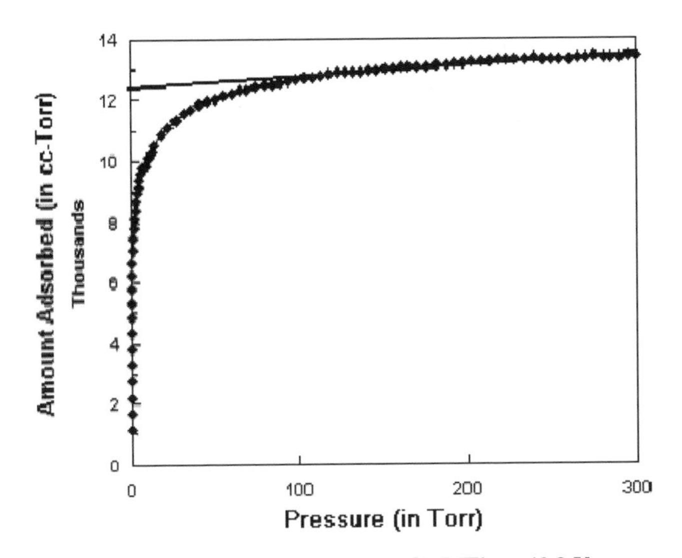

Figure 5. Adsorption isotherm of hydrogen on SWNT's at 40.2 K.

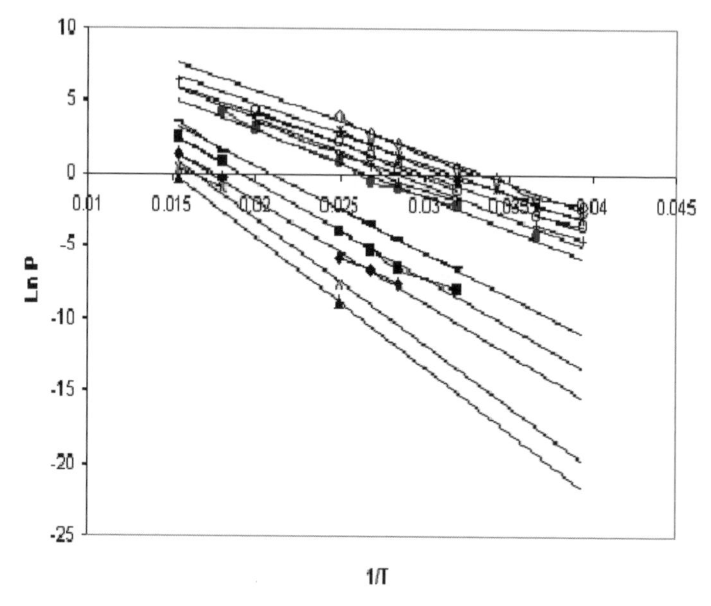

Figure 6. Ln P vs.1/T for coverages ranging between 1000 (lowest line) to 12,000 (topmost line) cc-Torr.

plotted LnP vs. 1/T for N different values of the fractional coverage. The resulting curves are shown in figure 6. The slopes of these straight lines multiplied with appropriate factor were used to find out the isosteric heats of adsorption at that particular fractional monolayer coverage.

The isosteric heat values were found to be a decreasing function of increasing fractional monolayer coverage. The value of isosteric heat of adsorption ranged from 80 mev at a fractional monolayer coverage of 0.08 to 38 mev near monolayer completion. Similar dependence of isosteric heats of adsorption with increasing coverage was also found in a very recent work for hydrogen and other adsorbates [12,16].

The values of isosteric heats obtained can be used to determine the binding energy. This value was found to range with in 72 mev for 0.08 monolayer coverage to 30 mev at monolayer completion. This implies that the lower coverage binding energy for hydrogen on as produced SWNT bundles is 1.73 times greater than that of the binding energy of hydrogen on graphite [14]. Similar increase in binding energies for Xenon [15] Methane [16] and Neon [15] was found for other adsorbates in some of our previous work.

To assign which adsorption sites correspond to the higher binding energy regions for hydrogen we performed monolayer adsorption isotherm measurements for xenon and argon on the same sample at 118 K and 82 K, respectively. These isotherms are shown in figures 7 a and b. The choice of axes in this figure is same as that of figure 4. The SSA obtained from the monolayer completion for these two gases were compared to the SSA obtained from hydrogen at monolayer completion. It has been observed in recent experimental studies that xenon [17,18] and argon [18] adsorb only on the outer grooves and the curved outer surface of the individual tubes on the periphery of the SWNT bundles. Therefore, if these two adsorbates yield similar SSA's than H_2, we can conclude that all three gases are adsorbing at similar sites. This is indeed the case, we obtain a SSA of

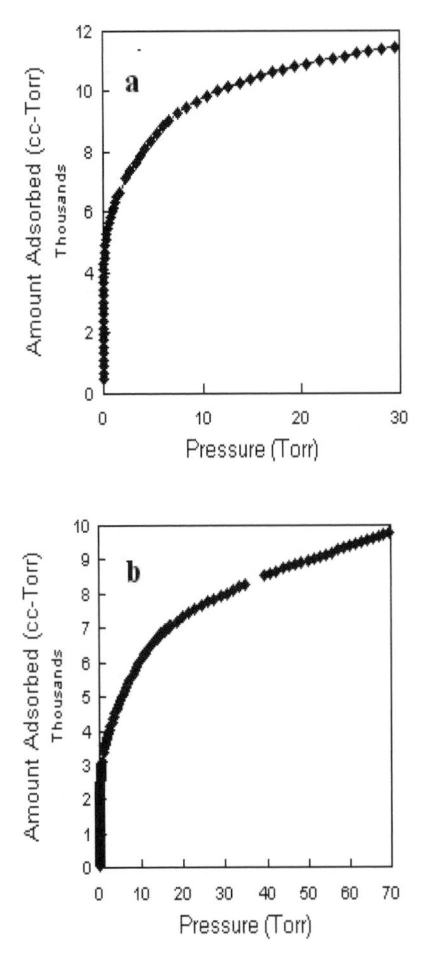

Figure 7. Adsorption Isotherm of (a) Argon and (b) Xenon at 82 K and 150 K respectively.

161.1m^2/g for Xe, 170.8 m^2/g for Ar, and 166.7 m^2/g for H_2. The similar values of the SSA indicates that H_2 adsorption on SWNT's mainly occurs on the grooves and the outside surface of tubes located at the periphery of the bundles.

CONCLUSION

We have determined the adsorptive capacity of hydrogen and its binding energy value on as produced SWNT's. The similar fractional increase in the binding energy values and the similar SSA available for adsorption for hydrogen and for other gases on SWNT's suggests that there is no hydrogen adsorption in the IC's on as produced tubes. These results are important because they suggest that as produced SWNT's are probably not suitable as efficient hydrogen storage medium at the present stage. Clearly more work in the area of production of different forms of carbon nanomaterials is needed to provide us with materials which are more efficient in storing hydrogen for fuel cell applications.

ACKNOWLEDGEMENTS

I would like to acknowledge the Link Foundation Energy Fellowship program for their support. I would also like to thank Prof. M.W.Cole and Prof. O.E.Vilches for a numerous illuminating discussions and for sharing their results prior to publication. This work was also supported by NSF through grant #DMR–0089713, the Petroleum Research Fund, and the Research Corporation.

REFERENCES

[1] A. C. Dillon and M. J. Heben, "Hydrogen storage using carbon adsorbents: past, present and future", *Appl.Phys.A* 72, 133–142 (2001).
[2] S. Iijima and T. Ichihashi, "Single shell carbon nanotube with 1NM diameter" *Nature* 363, 603–605 (1993).
[3] A. C. Dillon, K. M. Jones, T. A. Bekkedahl, C. H. Kiang, D. S. Bethune and M. J. Heben, "Storage of hydrogen in single walled carbon nanotubes", *Nature* 386–377 (1997).
[4] Calbi MM, Cole MW, Gatica SM, Bojan MJ, Stan G "Condensed phases of gases inside nanotube bundles" *Rev. Mod.Phys.*73, 4 (2001).
[5] M. S. Dresselhaus, K. A. Williams and P. C. Eklund "Hydrogen adsorption in carbon materials" *Mater.Res.Soc.Bull.* 24, 45 (1999).
[6] R. G. Ding, G. Q. Lu, Z. F. Yan and M. A. Wilson, "Recent advances in the

preparation and utilization of carbon nanotubes for hydrogen storage" *J. Nanosci. Nanotech,* 1, 7, (2001).
[7] H. Cheng, Q. Yang and C. Liu "Hydrogen storage in carbon nanotubes" *Carbon,* 39, 1447, (2001).
[8] V. Meregalli, M. Parinello, "Review of theoretical calculation of hydrogen storage in carbon-based materials", *Appl.Phys.A* 72, 143–146 (2001).
[9] C. Journet, W. X. Maser, P. Bernier, M. Lamy de la Chapelle, S. Lefrants, P. Deniards, R. Lee and J. E. Fischer, *Nature* 388, 756 (1997).
[10] A. Zangwill *"Physics at surfaces"* Cambridge University Press, 1998.
[11] C. M. Brown, T. Yildirim, D. A. Neumann, M. J. Heben, T. Gennett, A. C. Dillion, J. L. Allenman, and J. E. Fisher, "Quantum rotation of hydrogen in single walled carbon nanotubes", *Chem.Phys.Lett.* 329, 311 (2001).
[12] T. Wilson, A. Tyburski, M. R. Depies, O. E. Vilches, D. Becque, and M. Bienfait, "Adsorpton of H_2 and D_2 on carbon nanotubes bundles", *J. Low Temp. Phys. 126,* 403 (2002).
[13] K. A. Williams, B. K. Pradhan, P. C. Eklund, M. K. Kostov and M. W. Cole, "Raman spectroscopic investigation of H_2, HD and D_2 physisorped single walled carbon nanotubes" *Phys.Rev.Letts.* 88, 165502 (2002).
[14] G. Vidali, G. Ihm, H. Kim and M. Cole, "Potentials of physical adsorption" *Surf. Sci. Rep. 12,* 133–181 (1991).
[15] S. Talapatra, A. J. Zambano, S. E. Weber and A. D. Migone, "Gases do not adsorb on the closed end single walled nanotubes", *Phy.Rev.Lett.* 85,138 (2000).
[16] S. Talapatra and A. D. Migone "Methane adsorption on bundles of closed ended single walled nanotubes" *Phys. Rev. B.* 65, 045416 (2002).
[17] A. J. Zambano, S. Talapatra and A. D. Migone, "Binding energy and monolayer capacity of xenon on single walled nanotube bundles", *Phys. Rev. B.* 64, 075415 (2001).
[18] S. Talapatra and A. D. Migone, "Existence of novel quasi one-dimensional phases of atoms adsorbed on the exterior surface of closed ended single walled nanotube bundles." *Phys. Rev. Lett.* 87,201606 (2001).

PART II

SIMULATION-TRAINING

Aids for Training Real-World Spatial Knowledge using Virtual Environments

Andrew G. Brooks

Research Laboratory of Electronics

Massachusetts Institute of Technology

Cambridge, MA 02139

Research Advisor: Nat Durlach

ABSTRACT

Virtual Environment (VE) technology seems to be an ideal medium for training spatial knowledge of real world locations; environments can be constructed in a highly detailed fashion, incorporating photographic data from the actual space, and trainees are not limited by physical constraints on their movement and viewing capabilities, nor the need to cooperate with the regular inhabitants of the space. Examining the concept theoretically, VE training should be able to match or even surpass training performed in the real world.

However, VE-based spatial training has not yet lived up to its potential. Deficiencies in the technology to date give VE experiences paucity in certain critical areas that seem to impact training effectiveness. I believe that what is needed to counterbalance these shortcomings is a suite of improved training interfaces or aids, both supernormal devices rendered in the virtual world and physical mechanisms for the trainee to use in the real world to control their virtual training experience. In addition, tools are needed to make it easier to evaluate the results of VE training sessions, particularly the aspect of training transfer to the real world.

I therefore describe some aids that I have developed for the purpose of training spatial knowledge acquisition of complex real world environments. In the virtual domain, a three-dimensional map interface is described that can be used both for assisting navigation and localisation during training and for testing of training effectiveness after an instructional session. An experiment is reported which indicated that this interface allowed experts to report their spatial knowledge significantly more accurately than using a traditional real-world metric. In the physical domain, a novel, low-cost, small-footprint, unencumbering locomotion interface is described that I believe will contribute to improved path integration and thus more accurate mental model construction when used in a training context.

INTRODUCTION

Due to lack of real-world physical constraints, and the ability to give the trainee multiple viewing perspectives, Virtual Environment (VE) technology seems to be ideal for the training of spatial knowledge. For the simulation community, a particularly interesting case is the training of spatial knowledge of specific spaces that exist in the real world. Training individuals to have configurational knowledge of particular real world spaces, particularly with regard to critical missions where lives may be at stake, is desirable but frequently not feasible to perform in the real world spaces themselves due to cost, lack of access, or danger constraints. Illustrative examples include a hostage rescue mission in a disputed area, or a mission to rescue personnel trapped by fire in a building that is difficult to reserve for training purposes.

Although VEs would appear to be ideal for such training purposes, and some research has been done in this area (see a summary in [1]), to date the effectiveness of VE training in terms of training transfer has fallen short of that achievable in the real world. This is likely to stem, to a large extent, from the many areas in which the technology available for creating and rendering VEs is unable to deliver the sensory richness experienced during interactions with real environments and situations. However, it is also the case that our approach to VE development largely concentrates on reducing this gap in verisimilitude rather than isolating those particular components that are important in achieving effective training, and developing technologies to allow us to exploit the "supernormal" (facilities or attributes that are not present in the real world) capabilities of virtual environments. In particular, although much has been written on training transfer (for example, see [2]), relatively little attention has been given to the transfer effects of VE spatial training aids.

In this paper I discuss the development of two such aids; one a virtual device that performs both a training and an evaluation function within the VE, and the other a physical mechanism that allows VE trainees to better engage their vestibular system when navigating the environment, as well as reduce the cognitive demands associated with the locomotion task. Not only is it predicted that both of these items will lead to an increased rate of training transfer, but the development of an assessment mechanism for allowing VE-based spatial knowledge testing to predict real-world transfer would be a significant boon to the process of designing VE training aids specifically with transfer in mind.

WORLD IN MINIATURE

One of the most important navigational aids in large scale, real world spaces such as the outdoors is a map. However, in more complicated and three-dimensional large-scale real world spaces, such as a building, providing a trainee with a map becomes more difficult. A typical solution to the problem of constructing a 3–D map in the real world is to supply instead a collection of 2D slices, such as a set of floorplans. This has obvious deficiencies for highly complex spaces that may not lend themselves immediately to such regular horizontal slicing. In the virtual world, however, we are not limited to the physical constraints that make real 3–D maps difficult to manufacture and manipulate. Therefore, the principal virtual training aid I have been investigating is a device known as the World In Miniature, or WIM.

First described in [3], the WIM can be thought of as a 3–D map, but in actuality offers the user a large degree of flexibility in terms of viewpoint selection and ability to examine complex environmental detail. The WIM functions as an object in the world, so the user is able to bring it into and out of view, move, rotate and disassemble it according to the operation of hand-held, motion-tracked physical analogues. Whereas a map is an approximate representation of certain pre-chosen aspects of an area, the WIM's level of detail can be selected to be identical to that of the surrounding world. Also, the WIM contains a miniature avatar of the user, located and oriented within the WIM in a manner that matches the location and orientation of the trainee within the normal-sized "mother" virtual world.

While the WIM was originally proposed as a locomotion device, subsequent studies have indicated that users do not like to use it thus [4], so I have concentrated on it as a tool for representing spatial structure of the world. It is believed that the WIM will assist training transfer both in terms of increasing the accuracy and detail of the post-training mental model, and improving the ability of the trainee to use spatial information from the environment to achieve appropriate self-placement within the mental model.

One of the most important features of my WIM is the way in which it allows the trainee to access occluded detail, an unavoidable aspect of complex, three-dimensional, enclosed, interior spaces. Other WIM implementations have used automatic techniques to remove walls between the user's viewpoint and the miniature avatar within the WIM [3] or have suggested presenting only some portion of the overall world immediately surrounding the user as the WIM [4]. However, I wish trainees to be able to use the WIM to place local structure in the context of the overall configuration of the space, so for any moderately large or complex environment these techniques seem disadvantageous. These mechanisms are also quite strictly coupled to the sole task of allowing users a direct line of sight to their own location in the WIM, rather than providing them with a means of examining any section of the environment through the WIM. Therefore, the WIM described here has a specific user interface concerned with the distortion of the WIM in order to allow trainees to use the WIM to view the entirety of the space, rather than assigning some purely automatic disassembly beyond their control.

This interface has been named the "saw", which is a planar virtual cutter that is operated by the trainee with the hand not holding the WIM. When the saw is brought into contact with the WIM, all detail on the "cut" side is removed, revealing the internal contents. Once the desired dissection is achieved, the trainee is able to lock the saw into place until a new decomposition is desired. The saw can be turned on and off with a finger switch, allowing the trainee to rapidly toggle between full and cutaway views of the WIM, to maximise the ability to place local spatial information into a global context. Further options are also available to operators; another switch restricts the saw to 45 degree discrete cuts, which are the most commonly used, and users are able to scale both the WIM and their own avatar in order to select a view that is appropriate to the examination being performed. In addition, to save time during rough localisation queries, the trainee is able to rapidly locate his or her avatar within the WIM without using the saw to explicitly remove occluding detail, by toggling the display of transparent position crosshairs extending beyond the boundary of the WIM.

The tangible user interface side of the WIM system bears some reporting here. It seemed appropriate to incorporate a kinesthetic component that matches the access mode that would be applied to a small model in the real world. The motion tracker controlling the WIM is therefore attached to a manipulable physical representation that does not necessarily suggest the model in size, shape or weight but makes its position and orientation obvious by touch. Conversely, cutting the model with a plane is intuitively represented by simply using the plane of the trainee's hand to suggest the cutting plane. The saw, therefore, is implemented as a motion tracker attached to a glove worn by the trainee, allowing the trainee's hand orientation itself to be applied to that of the saw.

The WIM system will be used in experiments to be conducted with a photorealistic VE of the basement, mezzanine and first floor of the building containing our laboratory, Building 36 at MIT. The VE was created using extruded floorplans for the geometry and digital photoimages. In addition, the software written for controlling the WIM was used for an experiment regarding evaluation of the technology for spatial knowledge testing in the virtual world and verification of the user interface, which is described in the following section.

WIM MANIPULATION EXPERIMENT

Human spatial knowledge, or the mental model of a specific space, develops through a range of categories as the space is learnt. The initial form of spatial knowledge is landmark knowledge, the ability to recognise objects and locations within the space. Following the acquisition of landmark knowledge comes route knowledge, the ability to plan and navigate a path between landmarks. Finally, the ability to accurately represent and visualise the relative positions and arrangements of landmarks in space is known as configurational knowledge. It should be noted that the progression between these forms of spatial knowledge is not automatic; acquisition of configurational knowledge is an effortful process, and without a deliberate attempt to acquire this form of knowledge, spatial aptitude in a specific space may well remain at the route level indefinitely.

If research is to be conducted into the acquisition of configurational knowledge, then a method must exist to evaluate the degree and accuracy of configurational knowledge possessed by the inhabitant of a space. Methods for assessing landmark and route knowledge are relatively simple, as basic recognition and planning tasks can be given to the individual, but the separation and quantification of configurational knowledge has proved a somewhat thornier problem. The most traditional method of assessment of configurational knowledge has been to use egocentric (first-person) range and bearing estimation, either landmark-to-landmark or via the "projective convergence technique" which involves the estimation of the range and bearing of various occluded landmarks from three different sighting locations, from the results of which various metrics can be calculated [5].

Range and bearing estimation methods suffer from a significant intersubject variation in the ability to accurately estimate these quantities. In contrast, since a fundamental aspect of configurational knowledge is the ability to generate an exocentric (external view) perspective, a popular traditional method of assessing configurational knowledge is to require individuals being tested to recreate such exocentric views, such as sketching a map, *e.g.* [6,7]. However, map construction tasks are not without drawbacks of their own. For example, quantifying the accuracy of a sketched map is difficult, and the task relies heavily on subject skills unrelated to the specific configurational knowledge, such as drawing and layout ability. In addition, due to the nature of the typical drawing medium, map-sketching tasks do not scale easily to the third dimension. Efforts to combat the first two problems have made use of "pick-and-place" map reconstruction, requiring the subject under evaluation to move a collection of magnets, paper cutouts or similar physical representations into position on a board, and then scoring the result based on various position error measures [8–11]. Once again, however, these methods have to date concentrated on the assessment of configurational knowledge of spaces that can easily be reduced to two-dimensional representations.

For various reasons, the assessment of configurational knowledge about a space has traditionally been performed in the real world space. The majority of assessment tasks involve the quantification of configurational knowledge that was originally acquired in the real world anyway, and in the case of assessment tasks involving configurational knowledge acquired in a virtual environment, the primary concern of the overarching study has often been the transfer of training to the real world, naturally suggesting real-world evaluation techniques [1]. However, it is conceivable that VE technology, perhaps in parallel with the traditional methods, would provide an ideal framework for extending exocentric configurational knowledge tests into three dimensions—in particular, for predicting the transfer effectiveness of a particular type of training or VE-based training aid.

Therefore a new variant on the pick-and-place map construction task is proposed that uses non-immersive (no effort is made to obscure the real world from the senses of the operator) VE equipment. Test subjects will be required to place landmarks, using a three-dimensional position sensor, within the exterior boundaries of a space, visually represented on a computer screen. As the task will be completely performed within a software simulation, the locations of the placed objects will be able to be recorded with a high degree of accuracy. The experiment described in this section was designed with two aims in mind: first, to determine whether performance on this task was an appropriate predictor of real world configurational knowledge; and second, to determine whether the WIM interface was intuitive and simple enough for even novice users to be able to manipulate it accurately and effectively.

The experiment was a repeated measures design in which each member of a group of subjects would perform (in random order) the same landmark placement test via four different interfaces. Three of these were exocentric and computer-based using the non-immersive VE, the difference between them being the specific manner of exocentric views available to the subject, whereas the other was the "traditional" baseline task for such testing, the range and bearing estimation in the real world. The three exocentric viewpoints were as follows: fixed viewpoint with perspective, controllable viewpoint (the WIM condition), and a triplet of fixed orthographic projections of the space (front, right, overhead).

In the three VE conditions, the subjects moved a physical three degree-of-freedom position tracker in space, to control a virtual cursor in the model. The cursor was used to place the landmarks within the model shell. In addition to the kinesthetic feedback of moving the tracker, subjects received visual feedback in the form of a set of crosshairs. As this particular experiment was intended to evaluate the test method and interfaces themselves, rather than the actual configurational knowledge of the subjects, it was important to ensure that all test subjects be "experts" in the spatial layout of the area being tested. Several methods were therefore employed to confer on the subjects perfect knowledge of the locations of the landmarks, in order to prevent any training effects from being observed in the result data.

The baseline task used as a control, exocentric range and bearing estimation, involved each subject standing in a fixed position and indicating the directions and distances to other landmark objects as if viewing from the location of a particular landmark object. Recent work in the psychological literature indicates that humans perform much better at tasks involving mentally translating their own bodies than tasks in which they must perform mental rotations [12;13]. Therefore the axis of the test room was fixed in the real world, and thus each subject had only to make a mental translation to the location of each object, but was able to rotate physically to indicate the answer direction, obviating the need for error-prone mental rotations.

The subjects who participated in this experiment were 9 students, 8 of whom were current undergraduate and graduate students at the Massachusetts Institute of Technology. Subjects were aged 18–24; 6 were male and 3 female. All had prior experience performing tasks on computers using a video display monitor, and none reported any visual impairment. Subjects were not compensated for their participation in the experiment.

Subjects were verbally instructed that they were to perform a task that evaluated their ability to reconstruct the spatial arrangement of a set of landmarks. They were then shown the physical layout of the landmarks to be described, including being given a written description of their relative Cartesian positions and being placed in an immersive VE of the area to be tested. Subjects were instructed to use the immersive experience to further confirm the relative arrangement of the landmarks, and were allowed to remain in the VE as long as desired.

Subjects were then instructed in the use of each non-immersive VE placement option, and given time to practice moving the cursor. They were also trained in using the orientation tracker to perform pointing for the egocentric range and bearing estimation task. Following the test familiarisation phase, they were instructed to place the landmarks they had learned via each of the test methods, in random order.

The data from the VE-based exocentric testing methods consisted of a set of Cartesian object coordinates, so the errors in object placement were calculated according to the simple linear distance of the locations of the placed objects from the correct object locations. In the egocentric estimation case, the responses were treated as spherical coordinates, corrected for sensor offset and converted to Cartesian object locations, and the linear error distance computed as before. The mean performance across all the testing methods is shown in Figure 1. The ordinate axis shows the magnitude of the error, so higher values indicate worse performance.

The error data was tested for homogeneity of variance in order to determine its suitability for comparison via the ANOVA statistical test. The vari-

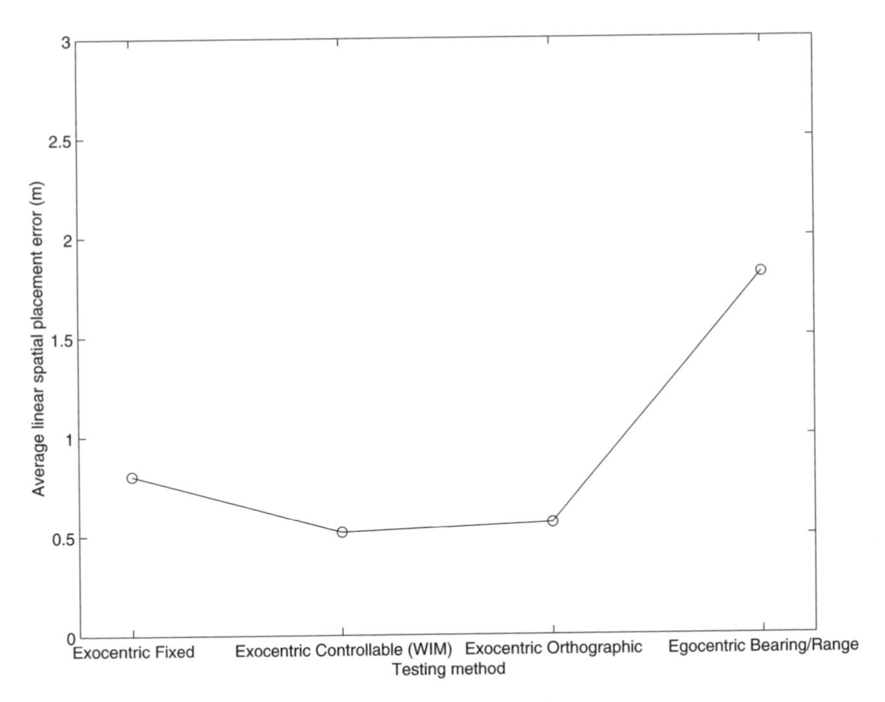

Figure 1. Mean performance according to test condition.

ance of the egocentric results demonstrated inhomogeneity with the variances of the exocentric tests, which were homogeneous with respect to each other. The egocentric results were therefore compared with the exocentric results using a paired t-test, showing a significant difference ($p << 0.001$). A one-way repeated measures ANOVA was performed on the exocentric data, indicating a significant difference in the samples ($F=6.659$, $p=0.0274$). Post hoc comparisons showed that the performance in the case of the single fixed viewpoint was significantly worse than the controllable or multiple orthographic viewpoints ($p=0.0283$) but that the differences between the controllable and multiple orthographic viewpoints were not significant. In all statistical tests, the significance threshold was preselected to be 0.05.

The results of this experiment demonstrate that exocentric methods for assessing spatial knowledge conclusively outperform the most common egocentric method, that of range and bearing estimation to landmarks. While this is no surprise, as the same conclusion has been reached concerning two-dimensional exocentric spatial knowledge testing methods, the results of this experiment also show that the particular method described here of extending these exocentric methods into three dimensions is valid despite the limitation of the display method, namely the use of a two-dimensional projection for representing three-dimensional data. Most importantly, the most accurate overall condition was the WIM condition, indicating that not only is the interface manageable for novice users, but that a high degree of accuracy is achievable. Though the WIM condition was not significantly better than the orthographic condition, the fact that it was shown to be comparable with such a multiply redundant condition indicates that problems should not be expected associated with cognitive overload of trainees using the WIM.

In summary, it seems evident that three-dimensional exocentric spatial arrangement tests using VE technology are effective at diagnosing the configurational knowledge possessed by expert subjects, and are more so than the traditional egocentric method of range and bearing estimation, even when subjects are able to mentally approximate the latter values from exact coordinates. Furthermore, the benefit conferred by allowing the selection of any viewpoint (via the WIM) when performing the exocentric task seems to outweigh any detraction in performance, or otherwise be unaffected, by the additional spatial processing required in the execution of the viewpoint selection subtask.

FOOTSTEPPER

One practical need that was identified for training in a large-scale virtual environment was a means of moving through the world in a more realistic manner than typical hand control or gesture based interfaces allow. It is a general problem in interface design that certain modalities become "overloaded" with many functions, requiring the user to keep track of functional modes or complicated codings to achieve actions that in the real world might require little thought. In addition to this problem, there is a tendency to design controls using the prevailing interface hardware even when this hardware (e.g., a keyboard or joystick) bears only an abstract correspondence to the desired action. These tendencies are understandable given the ubiquity of keyboards and joysticks, and the ease of integrating their output into a software system, but there are clear benefits to expanding the repertoire of interface devices to more closely match the capabilities of the human body.

In the present case, it is fairly obvious that, for most people, the preferred mode of moving about a large-scale (building sized) space is simple walking from place to place. A locomotion interface for a VE walkthrough should take advantage of this intuitive mode of action. There has been a great deal of work performed on locomotion interfaces appropriate to various scales of VE and tuned to different needs, including work performed at Sarcos, MERL, Virtual Space, and elsewhere. Two significant interfaces worth noting are bicycle-based, for use in outdoor, very large scale worlds: Ben Brown's Tektrix bike built for Georgia Tech's virtual recreation of the Atlanta Bicycle Road Race of the 1996 Olympics, and Mike Benjamin & Thatcher Ulrich's VR Bike input device for MERL Diamond Park. These interfaces are noteworthy because they show what can be accomplished by combining commercially available parts (like bicycles) with clever motorised attachments for simulating the forces that normally occur when moving through the world.

In the area of walking interfaces, the seminal work on the "Virtual Perambulator" [14] is particularly relevant in that using it requires effort as well as motion to move, just as is required in the real world. Relevant research on imparting constant and inertial forces to cause the user to expend energy has been carried out at the University of Utah using the Sarcos Treadport interface [15]. The possibility of physical exertion and fatigue when traversing a space is likely to add to the user's sense of presence by allowing the physiological state of the user in the VE to match that of the corresponding experience in the real world. Thus the user's mental state and physical state will be appropriate to the activity in the VE simulation. There may also be more subtle but important implications to using effort-based locomotion, involving spatial apperception, path integration and distance estimation. These aspects have not yet been thoroughly studied, but have been topics of investigation in our laboratory for the past few years beginning with our "Fingerwalker" system [16] through the present studies using the FootStepper.

In designing the FootStepper, three interface requirements were considered of primary importance; that the user be able to locomote in a straight line, turn in place, and reverse direction, all reliant solely upon the modalities of leg motion and body attitude. Walking forward using the FootStepper involves simply pumping the foot pads of the device symmetrically as one would expect. Turning in place is achieved by pumping the pads asymmetrically (as though dragging one foot), and walking backwards is facilitated by shifting one's weight to the rear of the foot pads while walking in the usual fashion. This important latter activity, with the weight concentrated on the heels, is strikingly similar to walking backwards in the real world, particularly when compared with other toggle methods of selecting "reverse gear."

The FootStepper is based on a sturdy but inexpensive exercise machine. This unit was originally selected not only for its robust construction, but also for the simple adjustment of step amplitude and the integral dampers. The dampers were first modified to a resistance appropriate for both walking and running behaviour by draining their silicone fluid reservoirs. A

Figure 2. Photograph of the FootStepper in use in a VE walkthrough of Building 36 at MIT.

future enhancement to the system will be to add electronically controlled valves to the damper in order to vary the damping factor under computer control (to simulate walking on hilly terrain).

The foot pedal position signals are derived from the action of a magnet (attached to the moving lever) on a linear Hall-effect sensor, feeding into a

Figure 3. Close-up photograph of the FootStepper, which is based on the StepFlex Compact Fitness Center Mini Stepper, sold at Brookstone.

microcontroller circuit that converts the sensor values into serial output. The heel pressure sensors were constructed using Zoflex ZF series pressure-sensitive conductive rubber from Zilor Inc., sandwiched between a pair of aluminium sheets. Upon processing by the microcontroller box, the heel sensors operate as a binary switch, detecting higher than normal pressure and registering reverse movement in the serial output. Due to the forward-tilted attitude of the footpads, test results did not indicate a necessity to recalibrate the sensors for normal variations in total body mass of the operator.

Software was then written to interface the FootStepper to a virtual environment walkthrough. The software consisted of a generic "motion model" for the user's viewpoint, layered on top of a serial communications and FootStepper querying and processing engine. The motion model essentially keeps track of two variables; the user's rate of linear motion (forward and backward) and the amount of rotational motion (turning left and right). The software continuously monitors the foot position data from the serial connection, evaluating local maxima and minima to determine the occurrence of 'step' events, and the heel sensors monitored to determine the direction of motion. The step events are then filtered in time and amplitude to compute a linear velocity and an estimate of the skew, for turning purposes. There is also a predictive element to the maximum and minimum computations, in order to facilitate the taking of single steps, an important form of small-scale maneuvering. A decelerator provides negative feedback to the step control system, allowing the user to come to a smooth halt when stepping ceases.

The FootStepper software currently supports only linear motion and turning in place as priority was given to preventing the registering of undesired turns, in order to make the interface as robust as possible for novice users, especially given the added task of maintaining balance while wearing a head-mounted display. In addition, there is an important latency present in the interface. Some latency is inevitable in any virtual environment locomotion interface, as the device must spend time recording and interpreting users' movements in the real world before updating their virtual location and scene view. The principal latency inherent in the FootStepper is that no update to the user's velocity and hence perception of motion is initiated until the first complete step has been recorded. This was necessary due to the use of a single modality for both position and orientation control—during the first moments of a step, the system has no way of determining whether the step is symmetric or asymmetric, and thus whether the user intends to move or turn. This latency does not seem overly distracting, and the reduced amount of cognitive overhead and coordination required through the reliance on a single modality led this trade-off to be considered acceptable.

EFFORT-BASED LOCOMOTION STUDY

In addition to the benefits in terms of operator intuitiveness and cognitive ease of use of a locomotion interface that exploits the natural walking modality, we hypothesise that an effort-based locomotion system such as the FootStepper will lead to a more tangible training benefit: better survey configurational knowledge development through improved path integration.

A number of studies have suggested that the incorporation of multiple sensory cues, particularly those involving some sort of correlation between physical effort and distance travelled, improves the ability of a human subject to interpret the environment being experienced. For example, experiments comparing navigation of a virtual maze using the "Virtual Motion Controller", a body-controller interface, with the same task using a joystick, show that configurational knowledge of the virtual space was significantly improved by using the body interface [17]. Moreover, the observed improvement increased with the complexity of the virtual maze. Similarly, studies of perceived geographic slant showed that subjects consitently overestimated the angle of hills and inclines under conditions that necessitated more personal effort in walking, such as encumbrance or fatigue [18].

On a more separable level, it has been demonstrated that path integration of rotational motion in a VE was significantly improved by allowing subjects to control their rotation by physically rotating themselves in an effort-based manner [19]. I have proposed a translational analogue of this experiment to assess the effect on linear path integration of our FootStepper effort-based locomotion system. I predict benefits for distance integration under translational movement when using this device similar to the enhanced rotational integration reported in [19] when subjects are allowed to rotate themselves rather than the environment.

The experimental description is as follows. The subject is placed in a reasonably large VE. A red target appears some distance in front of him or her. The subject is instructed to locomote to the target, either walking with the FootStepper or using a joystick, according to test condition. Upon reaching the target, the subject must turn around, either skew-walking to turn in place in the case of the Footstepper, or by rotating him or herself by way of a circular rail. A blue target now appears some distance back in the direction from which he or she approached, and the subject must locomote back to it. This last task repeats for a number of blue targets. Finally, small green targets appear at equally spaced locations, and the subject is asked to locomote to the one that is in the position in the virtual world of the original red target.

Although the task requires turning in place, the quantity to be integrated by the subjects is solely a one-dimensional translational distance. The accuracy of the subject's path integration is indicated by the distance between the green marker representing his or her final position and the actual location in the VE of the desired destination. As these distance values can be thought of as the number of markers' difference between the selected and correct positions, the errors are discrete and immune to noise associated with users' difficulties in precise self-positioning within the VE. This experiment is not yet complete, so no results can be reported at this time.

FUTURE WORK

In addition to the experiments mentioned in the text that are yet to be commenced or completed, such as verification of improved spatial training transfer from use of the WIM training aid, and improved linear path integration from use of the FootStepper interface, a number of other training aids have been considered for development and evaluation. As with the ones detailed here, these aids are contemplated with configurational knowledge in mind, and intended to be tested with the WIM-based landmark placement test as well as traditional measures in the real world.

Taking inspiration from the redundant navigational cues frequently used inside large buildings to assist in self-placement, one avenue for future training aid experimentation in the visual domain is through the use of unobtrusive visual overlays that convey certain information to the trainee without interfering with the viewing of environmental detail. Some examples are the use of colour-based feedback for position or orientation, and the placement of spatial reference information in some intuitive format on a "heads-up display".

In addition to visual training aids, future work will involve experimentation with the inclusion of spatial information conduits operating in the other sensory arenas. In particular, I believe that supernormal spatialised audio can be used to create a training aid that will achieve positive transfer, in addition to the value of aural stimulation as a means of replicating real-world detail. Audible sound can provide a way to inform the trainee of the location of specific regions beyond visual occlusion; for example, proximity and direction to a source of noise (such as a machine room or elevator shaft) can allow a navigator to connect local pieces of the developing mental model into a global structure even though their interconnection is not visible from the first person viewpoint.

I propose generating artificial noise sources at strategic places throughout the VE for this purpose. As these "soundmarks" would contribute to the generation of a more accurate mental model in terms of configurational knowledge of the space, I believe that the resulting model will transfer despite the absence of identical stimuli in the corresponding real environment. Furthermore, since the presence of audible cues does not provide any visual distraction, I believe the ability to use the available visual information to self-localise within the mental model will not be impaired.

Finally, it is hoped that the empirically observed transfer characteristics of the collection of VE training aids that are developed will assist in the construction of a predictive model of training transfer. This model should be able to assist in the design of future VE spatial training aids by allowing the designer to isolate those components of an aid that maximise positive transfer, and ensure that they are included and not diluted or distracted from in the final device.

ACKNOWLEDGEMENTS

Development of the FootStepper was performed as a collaboration with Dr Thomas E. von Wiegand at MIT, so my gratitude is extended to him as well as to my advisor, Nat Durlach. In addition to the Link Foundation, the work described here was performed with support from ONR grants N00014–96–1–0379 and N00014–01–1–0062, as well as Air Force AASERT F49620–98–1–0442.

REFERENCES

[1] G. Koh, T.E. von Wiegand, R.L. Garnett, N.I. Durlach and B. Shinn-Cunningham, "Use of Virtual Environments for Acquiring Configurational Knowledge about Specific Real-World Spaces: Preliminary Experiment", *Presence: Teleoperators and Virtual Environments*, 8 (6), 632–656 (1999).

[2] K.E. Williams, "The Role of Cognitive Simulation Models in the Design of Effective Training and Instructional Systems", unpublished report (1997).

[3] R. Stoakley, M.J. Conway, R. Pausch, "Virtual Reality on a WIM: Interactive Worlds in Miniature", *Proceedings of CHI '95: Mosaic of Creativity*, 265–272 (1995).

[4] R. Pausch, T. Burnette, D. Brockway and M.E. Weiblen, "Navigation and Locomotion in Virtual Worlds via Flight into Hand-Held Miniatures", *ACM Transactions on Graphics* (1995).

[5] A.W. Siegel, "The externalization of cognitive maps by children and adults: In search of ways to ask better questions", in L.S. Liben, A.H. Patterson and N. Newcombe (eds.), *Spatial representation and behavior across the life span: Theory and application*, 167–194, Academic Press (1981).

[6] K. Lynch, "The image of a city", MIT Press (1960).

[7] D. Appleyard, "Styles and methods of structuring a city", *Environment and Behaviour*, 2, 110–118 (1970).

[8] R.C. Sherman, J. Croxton and J. Giovanatto, "Investigating cognitive representations of spatial relationships", *Environment and Behaviour*, 11 (2), 209–226 (1979).

[9] J.C. Baird and M. Wagner, "Modeling the creation of cognitive maps", in H. Pick and L. Acredolo (eds.), *Spatial Orientation*, Plenum, 321–366 (1983).

[10] C.D. Wickens and T.T. Prevett, "Exploring the dimensions of egocentricity in aircraft navigation displays", *Journal of Experimental Psychology Applied*, 1 (2), 110–135 (1995).

[11] D.A. Waller, "An assessment of individual differences in spatial knowledge of real and virtual environments", unpublished doctoral dissertation, University of Washington, Seattle (1999).

[12] J.J. Rieser, "Access to knowledge of spatial structure at novel points of observation", *Journal of Experimental Psychology: Learning, Memory and Cognition*, 15 (6), 1157–1165 (1989).

[13] C.C. Presson and D.R. Montello, "Updating after rotational and translational body movements: Coordinate structure of perspective space", *Perception*, 23, 1447–1455 (1994).

[14] H. Iwata and T. Fujii, "Virtual Perambulator: A Novel Interface Device for Locomotion in Virtual Environment", *IEEE Proceedings of VRAIS '96*, 60–65 (1996).

[15] R. Christensen, J.M. Hollerbach, Y. Xu and S. Meek, "Inertial force feedback for the Treadport locomotion interface", *Presence: Teleoperators and Virtual Environments* 9 (1), 1–14 (2000).

[16] S.B. Fitch, "The Finger Walker: A Method to Navigate Virtual Environments", M.Eng. Thesis, Massachusetts Institute of Technology, Cambridge (1998).

[17] B. Peterson, M. Wells, T. Furness and E. Hunt, "The Effects of the Interface on Navigation in Virtual Environments", *Proceedings of the Human Factors and Ergonomics Society 42nd Annual Meeting*, 1496–1500 (1998).

[18] M. Bhalla and D.R. Proffitt, "Visual-Motor recalibration in geographical slant perception", *Journal of Experimental Psychology: Human Perception & Performance* 25 (4), 1076–1096 (1999).

[19] S.S. Chance, F. Gaunet, A.C. Beall and J.M. Loomis, "Locomotion mode affects the updating of objects encountered during travel: The contribution of vestibular and proprioceptive inputs to path integration", *Presence: Teleoperators and Virtual Environments* 7, 168–178 (1998).

Light Field Mapping: Efficient Representation of Surface Light Fields

Wei-Chao Chen

Department of Computer Science
University of North Carolina at Chapel Hill
Chapel Hill, NC 27599-3175
Research Advisor: Henry Fuchs

ABSTRACT

Recent developments in image-based modeling and rendering provide significant advantages over traditional image synthesis processes, including improved realism, simple representation and automatic content creation. Representations such as Plenoptic Modeling, Light Field, and the Lumigraph are well suited for storing view-dependent radiance information for static scenes and objects. Unfortunately, these representations have much higher storage requirement than traditional approaches, and the acquisition process demands very dense sampling of radiance data. With the assist of geometric information, the sampling density of image-based representations can be greatly reduced, and the radiance data can potentially be represented more compactly. One such parameterization, called Surface Light Field, offers natural and intuitive description of the complex radiance data. However, issues including encoding and rendering efficiency present significant challenges to its practical application.

We present a method for efficient representation and interactive visualization of surface light fields. We propose to partition the radiance data over elementary surface primitives and to approximate each partitioned data by a small set of lower-dimensional discrete functions. By utilizing graphics hardware features, the proposed rendering algorithm decodes directly from this compact representation at interactive frame rates on a personal computer. Since the approximations are represented as texture maps, we refer to the proposed method as Light Field Mapping. The approximations can be further compressed using standard image compression techniques leading to extremely compact data sets that are up to four orders of magnitude smaller than the uncompressed light field data. We demonstrate the proposed representation through a variety of non-trivial physical and synthetic scenes and objects scanned through acquisition systems designed for capturing both small and large-scale scenes.

INTRODUCTION

In the field of computer graphics research, the quest for photo-realistic image synthesis has traditionally focused on the study of light transport mechanisms. Starting with analytical models of both the surface geometry and reflectance properties, these algorithms render images by using a combination of physical simulations and heuristics. Following this paradigm, we have witnessed a tremendous improvement in image quality, rendering efficiency and scene complexity over the past three decades. On the other hand, despite the continual improvement of modeling tools and techniques, model creation remains labor-intensive, and it has become the primary bottleneck in the traditional paradigm.

The recent proliferation of inexpensive but powerful graphics hardware and new advances in digital imaging technology are enabling novel methods for realistic modeling of the appearance of physical objects. On the one hand, we see a tendency to represent complex analytic reflectance models with their sample-based approximations that can be evaluated efficiently using new graphics hardware features [1–3]. On the other hand, we are witnessing the emergence of Image-Based Rendering and Modeling (IBRM) techniques [4–7] that attempt to represent the discrete radiance data directly in the sample-based format without resorting to the analytic models at all. These techniques are popular because they promise a simple acquisition and an accurate portrayal of the physical world. The approach presented here combines these two trends. Similar to other image-based methods, our approach produces a sample-based representation of the surface light field data. Additionally, the proposed representation can be evaluated efficiently with the support of existing graphics hardware.

Light Field Mapping Overview

A combination of synthetic and physical objects rendered using Light Field Mapping is shown in figure 1. A surface light field [8–10] is a 4–dimensional function $f(r,s,\theta,\phi)$ that completely defines the radiance of every point on the scene surface geometry in every viewing direction. The first pair of parameters of this function (r,s) describes the surface location and the second pair of parameters (θ,ϕ) describes the viewing direction. In practice, a surface light field function is normally stored in sampled form, where as the geometry information are normally represented as surface mesh. Figure 2 illustrates the surface light field parameterization.

Because of its large size, a direct representation and manipulation of the light field data is impractical. We propose to approximate the discrete 4–dimensional surface light field function $f(\cdot)$ as a sum of products of lower-dimensional functions

Figure 1. A combination of synthetic and physical objects rendered using Light Field Mapping. Complex, physically realistic reflectance properties of this scene are represented and visualized.

$$f(r,s,\theta,\phi) \approx \sum_{k=1}^{K} g_k(r,s)\ h_k(\theta,\phi). \tag{1}$$

We demonstrate that it is possible to construct approximations of this form that are both compact and accurate by taking advantage of the spatial coherence of the surface light fields. This is accomplished by partitioning the surface light field data across small surface primitives and building the approximations for each part independently. The proposed partitioning also ensures continuous approximations across the neighboring surface elements. By taking advantage of existing hardware support for texture mapping and composition, we can visualize surface light fields directly from the proposed representation at highly interactive frame rates. Because the discrete functions g_k and h_k encode the light field data and are stored in a sampled form as texture maps, we call them the *Light Field Maps*. Similarly, we refer to the process of rendering from this approximation as *Light Field Mapping* [10].

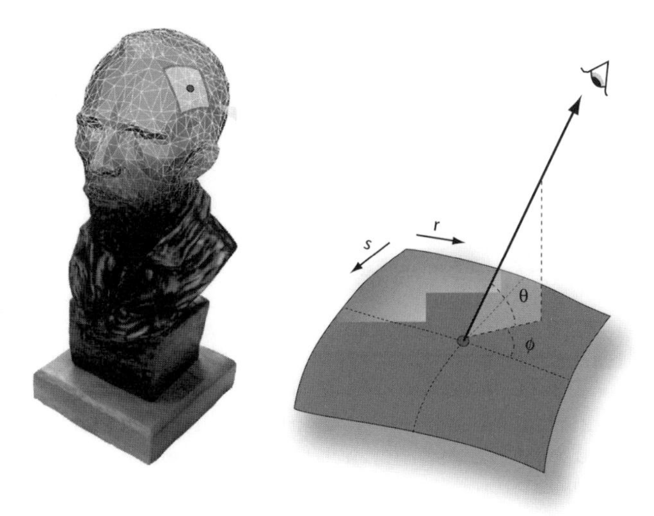

Figure 2. The surface light field parameterization. This parameterization is suitable for representing static sample-based scenes.

We make the following contributions to analysis and representation of image-based data and to hardware-accelerated rendering of image-based models.

- Partitioning of IBRM radiance samples on geometry for high-quality compression—We propose a novel type of light field data partitioning which allows efficient compression without introducing discontinuity artifacts.
- Efficient and high-quality compression algorithms for the radiance samples—We propose a class of compression algorithms that use simple linear approximations and are entirely data-driven. These algorithms work very well for intricate real-life scenes.
- A simple rendering algorithm suitable for real-time visualization—The rendering algorithm decompresses on-the-fly and renders using commodity graphics hardware at highly interactive rates.

In a later section, we introduce the proposed partitioning and approximation framework. Then we propose efficient rendering algorithms that allow on-the-fly decompression in graphics hardware. This representation are stored as images and can be further compressed using image processing algorithms, and several algorithms we have experimented with are presented. A description and implementation of the acquisition system used in this research is provided.

RELATED WORK

Reflection Models. Much conventional graphics research represents the surface reflectance properties as a model in the form of 4–dimensional Bidirectional Reflectance Distribution Function (BRDF) $r(u_i, f_i, u_o, f_o)$. A BRDF defines the ratio of the outgoing radiance at direction (u_o, f_o) to the incoming irradiance from direction (u_i, f_i). Many earlier methods represent the BRDF analytically. These methods can be further classified into two categories, namely empirical models [11–15] and physically-based models [16–19].

A recent research trend is to approximate BRDF by lower-dimensional sampled functions to facilitate hardware-accelerated rendering. Heidrich *et al.* [1] investigated and concluded many analytical BRDFs can be separated into products of 2–dimensional functions. Kautz and McCool [2] propose hardware-assisted rendering of arbitrary BRDFs through their decomposition into a sum of 2D separable functions. The homomorphic factorization of McCool *et al.* [20] generates a BRDF factorization with positive factors only, which are easier and faster to render on the current graphics hardware. Although the homeomorphic factorization framework can potentially support progressive encoding, the authors only presented an algorithm that limits the factorization to three factors. The research presented in this thesis is in part inspired by these research, although our application is fundamentally different. These methods are limited to sample-based representation of shift-invariant reflectance models. Our research focuses on complex, real world surfaces that may have different reflectance properties on each point of the surfaces. We also present a novel method for factorization of light field data that produces only positive factors [21]. The proposed method is significantly easier to implement and more accurate than the homomorphic factorization in [20].

To accommodate for spatial variance on the surface, there are also research efforts to extend the BRDF beyond 4 dimensions. For example, a Bidirectional Texture Function (BTF) is a 6–dimensional function that provides a BRDF for each 2–dimensional surface point [22–24]. Instead of explicitly sampling the BTF, Lensch *et al.* [25] took a different approach by taking a sparse set of photographs around an object and reconstructing a set of basis BRDFs from these photographs. Although this algorithm requires only a small set of photographs, it does not work well for complicated surfaces with many different BRDFs.

Image-Based Representations. Image-based methods synthesize novel images directly from input photographs. Some of the IBRM representations contain only samples from the acquisition process, such as the plenoptic modeling [4] and related image warping research [26–28], the

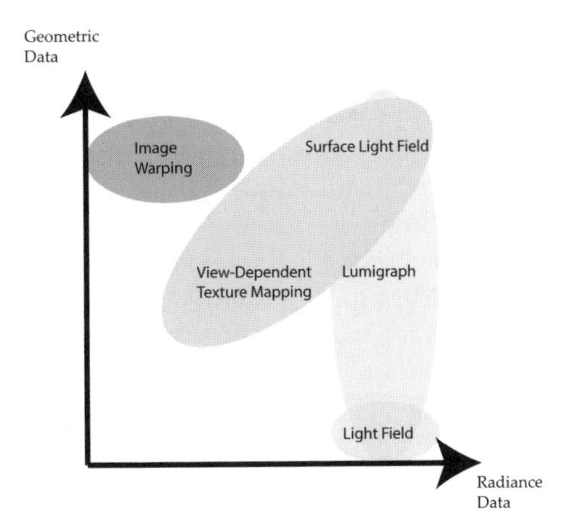

Figure 3. Taxonomy of image-based representations according to the amount of required geometric and radiance data.

light field rendering [5], and the lumigraph [6]. Others use traditional surface primitives to store geometric information, such as View-Dependent Texture Mapping (VDTM) [29–31, 7]. Figure 3 categorize these techniques according to the amount of geometric and radiance data required for each representation.

Chai *et al.* [32] pointed out that geometry information can be used to trade off radiance information. Therefore, some of the representations are more redundant than others, and can be potentially represented more compactly. Much research has been done on compression of light field data. Levoy and Hanrahan [5] use the Vector Quantization (VQ) technique [33] for compression of light field data. Magnor, Eisert and Girod have developed a series of disparity-compensated light field codecs [34–36]. Most of the research, however, do not lead to hardware-accelerated rendering algorithms.

VDTM and The Surface Light Field. View-Dependent Texture Mapping (VDTM) extends classic texture mapping algorithms by stores multiple images per surface primitive. During rendering, the algorithm texture-maps the surfaces using images taken from similar viewing directions. The primary advantages of VDTM techniques are that the required amount of radiance data in a VDTM is generally much lower than purely sample-based IBRM representations. However, as the number of input images increases, the amount of radiance data grows proportionally, and therefore VDTM algorithms does not scale to the number of input images.

Nishino *et al.* [37] proposed the eigen-texture method that compresses generalized VDTM data using Principal Component Analysis (PCA) algorithm [38]. They achieve approximately 20:1 compression ratio. Our proposed Light Field Mapping technique is sometimes confused with the eigen-texture method. The original formulation of the eigen-texture method only allows synthesis of input images and novel views on the path connected by a pair of images. This technique is thus not a general VDTM representation. Unlike for Light Field Mapping method, there are no reported real-time rendering algorithms for the eigen-texture representation.

The surface light field parameterization [8,9] is very similar to a VDTM parameterization. However, VDTM defines the viewing parameters θ,ϕ on a per surface primitive basis, whereas surface light field parameterization treats each surface point differently. VDTM is therefore an approximation of surface light field. Miller *et al.* in [8] proposed a method of rendering surface light fields from input images compressed using JPEG-like compression. Wood *et al.* [9] uses a generalization of VQ and PCA to compress surface light field and proposes a two-pass rendering algorithm that displays compressed light fields at interactive frame rates. Compared to this technique, we reported approximately two orders of magnitude faster rendering speed using Light Field Mapping approach in [39].

SURFACE LIGHT FIELDS APPROXIMATION

This section describes our method for approximating the radiance data. We first present a novel partitioning method that allows each partition to be approximated independently without introducing discontinuity artifacts. Then, we describe our approximation framework based on matrix factorization and decomposition algorithms. These approximations can be decoded and visualized very efficiently using the proposed Light Field Mapping rendering algorithms in a later section.

Surface Light Field Partitioning

A IBRM sample database is generally very large, and in practice, we can only handle a local scope of data during preprocessing. To enable efficient compression, the samples within a local scope should be highly coherent. For surface light field parameterization, surface primitives naturally define the unit of scope for our purposes. The units together form a partitioning of the sample database, and if we can process each part independently without introducing artifacts, we can parallelize both the approximation and decoding algorithms. Based on these observations, an effective surface light field partitioning scheme should should divide surface light field data. Furthermore, independent approximation of each part should not introduce artifacts.

Since the geometry of our models is represented as a triangular mesh, an obvious partitioning of the light field function $f(r,s,\theta,\phi)$ is to split it between individual triangles

$$f^{\Delta_t}(r,s,\theta,\phi) = \Pi^{\Delta_t}(r,s) f(r,s,\theta,\phi) , \qquad (2)$$

where $\Pi^{\Delta t}(r,s)$ is a step function that is equal to one within the triangle Δ_t and zero elsewhere. Because the partitioning breaks the original surface light field function on the triangle boundaries, we refer to this approach as *triangle-centered* partitioning. Unfortunately, when each function is approximated independently, the approximation process results in visible discontinuities at the edges of the triangles.

To eliminate the discontinuities across triangle boundaries, we propose to partition surface light field data around each vertex. The part of surface light field corresponding to each vertex is referred to as the *vertex light field* and for vertex v_j it is denoted as $f^{v_j}(r,s,\theta,\phi)$. This partitioning is computed by multiplying weighting to the surface light field function

$$f^{v_j}(r,s,\theta,\phi) = \Lambda^{v_j}(r,s) f(r,s,\theta,\phi) , \qquad (3)$$

where Λ^{v_j} is the barycentric weight of each point in the ring of triangles centered around vertex v_j. The value of Λ^{v_j} is equal to 1 on vertex v_j, and it decreases linearly toward zero at the boundary. Because of their shape, the weighting functions are often referred to as the *hat functions.* In Figure 4, the top row shows hat functions Λ^{v_1}, Λ^{v_2}, Λ^{v_3} for three vertices v_1, v_2, v_3 of

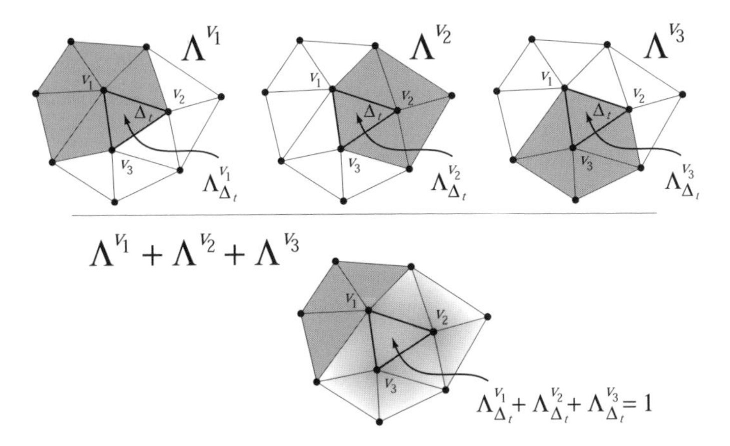

Figure 4. The finite support of the hat functions Λ^{v_j} around vertex v_j, j=1,2,3. $\Lambda^{v_j}_{\Delta t}$ denotes the portion of Λ^{v_j} that corresponds to triangle Δ_t. Functions Λ^{v_1}, Λ^{v_2} and Λ^{v_3} add up to one inside Δ_t.

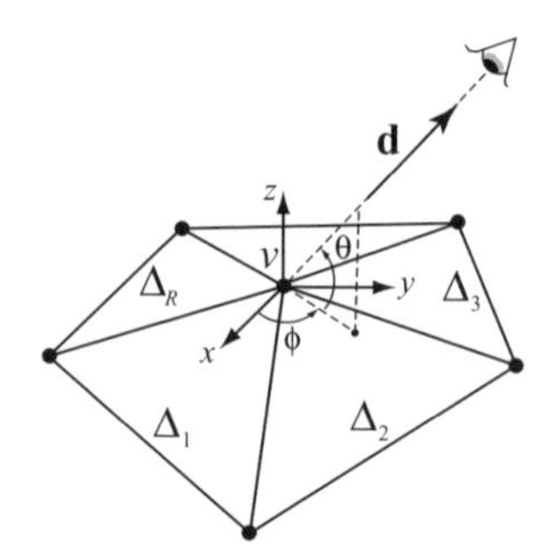

Figure 5. Local viewing angles (θ,ϕ) are the azimuth and polar angles of vector **d** in the reference frame (x,y,z). For vertex-centered approximation, the reference frame is attached at each vertex v where the z axis is parallel to the surface normal at the vertex.

triangle Δ_t. The bottom row of the same figure shows that these three hat functions add up to unity inside triangle Δ_t. Therefore, Equation (3) defines a valid surface light field partitioning, because the original surface light field can be reconstructed by simply summing up individual vertex light fields.

The final step of vertex-centered partitioning reparameterizes each vertex light field to the local vertex reference frame, as shown in Figure 5. A vertex reference frame is defined such that its z-axis is parallel to the normal at the vertex. The reparameterized (θ, ϕ) are simply the polar and azimuth angles of viewing directions in this frame. The vertex light field functions together with their corresponding local coordinates allow us to reconstruct the original data unambiguously. In the rest of the section, when we refer to a vertex light field function, we assume it to be expressed in the local coordinate, and for simplicity we use the same notation for both local and global parameters.

Vertex Light Field Approximation

As stated, vertex-centered partitioning of light field data allows us to approximate each partition independently without introducing discontinuity artifacts. We propose to approximate vertex light field as

$$f^{v_j}(r,s,\theta,\phi) \approx \sum_{k=1}^{K} g_k^{v_j}(r,s)\, h_k^{v_j}(\theta,\phi)\,. \tag{4}$$

The 2D functions $g_k^{v_j}$ contain only the surface parameters, and we refer to them as *surface maps*. Similarly, we refer to $h_k^{v_j}$ as *view maps*. The above approximation can effectively compress the function f^{v_j} if we only need a few approximation terms K to achieve high quality of approximation. We leverage existing matrix factorization algorithms to calculate the above approximations numerically. Before we discuss details of these algorithms, we describe how the vertex light field approximation problem can be transformed into a 2D matrix factorization problem.

For practical purposes, we assume that the vertex light field is stored in discrete format $f^{v_j}[r_p,s_p,\theta_q,\phi_q]$, where index p = 1,...,M refers to the discrete values $[r_p,s_p]$ describing the surface location within triangle ring of vertex v_j, and index q = 1,...,N refers to the discrete values $[\theta_q,\phi_q]$ of the viewing angles. We may rearrange the discrete vertex light field into a 2D matrix

$$\mathbf{F}^{v_j} = \begin{bmatrix} f^{v_j}[r_1,s_1,\theta_1,\phi_1] & \cdots & f^{v_j}[r_1,s_1,\theta_N,\phi_N] \\ \vdots & \ddots & \vdots \\ f^{v_j}[r_M,s_M,\theta_1,\phi_1] & \cdots & f^{v_j}[r_M,s_M,\theta_N,\phi_N] \end{bmatrix}, \tag{5}$$

where M is the total number of surface locations and N is the total number of views at each surface location. We refer to matrix $\mathbf{F}^{v_j}$ as the *vertex light field matrix.* In practice, to obtain vertex light field matrices from input images, we need to resample the input samples or photographs. Our 4D resampling algorithm are described later.

Matrix factorization algorithms construct approximate factorizations of the form

$$\tilde{\mathbf{F}}^{v_j} = \sum_{k=1}^{K} \mathbf{u}_k \mathbf{v}_k^T, \tag{6}$$

where $\mathbf{u}_k$ is a vectorized representation of discrete surface map $g_k^{v_j}[r_p,s_p]$ and $\mathbf{v}_k$ is a vectorized representation of discrete view map $h_k^{v_j}[\theta_q,\phi_q]$. The matrix F^{v_j} contains $M \times N$ samples, whereas its approximation contains $K \times (M + N)$. If $K << min(M,N)$, the size of approximation become much smaller than the size of original matrix $\mathbf{F}^{v_j}$.

Matrix Approximation Algorithms

In Equation 6, we observed that matrix factorization algorithms can be used to approximate the light field matrix. Although the light field matrices are generally quite large, computing full matrix factorization may be very time consuming. However, because we only need approximate factorization of the matrix, with proper implementation, the efficiency of these algorithms can be drastically improved.

We experimented with two algorithms to calculate the approximations: Principal Component Analysis (PCA) [38] and Non-Negative Matrix Factorization (NMF) [21]. Both of these algorithms compute matrix factorization in a form similar to Equation 6, and they have been used in a wide range of applications such as data compression and unsupervised learning. The differences between the two algorithms arise from the constraints imposed on the approximation factors $\mathbf{u}_k$ and $\mathbf{v}_k$. PCA enforces the factors $\mathbf{u}_k$ and $\mathbf{v}_k$ to be orthogonal vectors and keeps the factorization progressive; that is, once an order K factorization is computed, the first (K-1) pairs of vectors provide the best order (K-1) approximation. NMF, on the other hand, enforces all entries in vectors $\mathbf{u}_k$ and $\mathbf{v}_k$ to be positive. Unlike PCA, NMF produces a non-progressive factorization. In other words, a new approximation has to be recomputed when a different order K is chosen.

PCA Algorithm. The PCA factorization is based on computing the partial Singular Value Decomposition (SVD) of a matrix. The SVD of a $M \times N$ matrix $\mathbf{F}$ has the following form

$$\mathbf{F} = \sum_{i=1}^{min(M,N)} \mathbf{u}_i s_i \mathbf{v}_i^T, \tag{7}$$

where the column vectors $\mathbf{u}_i$ and $\mathbf{v}_i$ are the orthonormal left- and right-singular vectors of the matrix $\mathbf{F}$, respectively. The singular values s_i are ranked in non-ascending order so that $s_i \leq s_j$ for all $i < j$. Therefore, a partial sum of Equation 7 also gives the RMS optimal approximation to matrix $\mathbf{F}$ for the given dimensionality.

Because our goal is to approximate a matrix, it is unnecessary to perform the time-consuming SVD. In order to efficiently compute a K-term approximation of a matrix $\mathbf{F}$, we can instead compute the eigenvectors corresponding to the K largest eigenvalues of the covariance matrix $\mathbf{F}^T\mathbf{F}$. The power iteration algorithm is well suited to achieve this goal [40].

PCA algorithm requires the elements in matrix $\mathbf{F}$ to have zero mean. To satisfy this precondition, we subtract the average column vectors from the matrix $\mathbf{F}$ before performing PCA. The extracted vectors contain only surface parameters and can be treated as traditional diffuse texture maps during the rendering process.

NMF algorithm. We apply the iteratively algorithm presented by Lee *et al.* [21] to compute NMF approximation. Unlike PCA algorithm, all approximation vectors $\mathbf{u}_k$ and $\mathbf{v}_k$, $k = 1 \cdots K$ are updated simultaneously in every iteration, and we denote the matrix form of these vectors as $\mathbf{U}$ and $\mathbf{V}$ respectively. To improve approximation precision, we subtract the minimum column vector from matrix **F** before performing NMF. These minimum vectors are also treated as diffuse texture maps in the rendering process.

RENDERING ALGORITHMS

Earlier we described the process of partitioning and approximating surface light field data. The process generates sets of images collectively referred to as light field maps. In this Section we propose algorithms that use light field maps to achieve real-time rendering and on-the-fly decompression by taking advantage of graphics hardware features.

Rendering by Texture-Mapping

Let $g_k^{v_j}[r_p,s_p]$ be the surface map and $h_k^{v_j}[\theta_q,\phi_q]$ be the view map corresponding to the *k-th* approximation term of vertex light field $f^{v_j}[r_p,s_p,\theta_q,\phi_q]$. The approximation of the light field data for triangle Δ_t can be written as

$$\widetilde{f}^{\Delta_t}[r_p,s_p,\theta_q,\phi_q] = \sum_{k=1}^{K}\left(\sum_{j=1}^{3}(g_k^{v_j}[r_p,s_p]_{\Delta_t})(h_k^{v_j}[\theta_q,\phi_q])\right), \tag{8}$$

where index j runs over the three vertices of triangle Δ_t, and $g_k^{v_j}[r,s]_{\Delta t}$ denotes the portion of the surface map $g_k^{v_j}$ corresponding to triangle Δ_t. Equation (8) suggests that even though the approximation is done in a vertex-centered fashion, an approximation term for each triangle can be expressed independently as a sum of its 3 vertex light fields. This allows us to write a very efficient rendering routine that repeats the same sequence of operations for each mesh triangle. We now describe the rendering algorithm for one approximation term for one triangle. For simplicity we also drop the index k in functions g_k and h_k. We may devise a straightforward rendering algorithm as follows. For each target pixel in the triangle, we calculate the parameters (r,s,θ,ϕ) for this pixel, and evaluate Equation (8) to decode the approximation. This rendering algorithm is straightforward but it fails to exploit the inherent parallelism in the light field maps format.

To speed up the rendering process, we may utilize texture mapping features in contemporary graphics hardware. Texture coordinates within surface primitives are normally linearly interpolated from the vertex texture coordinates. Therefore, we need to transform the parameters of view maps into a space compatible with hardware texture coordinates interpolation. Assume vector **d** is the normalized viewing direction, and that vectors **x** and **y** correspond to the axes of the local reference frame. We may then

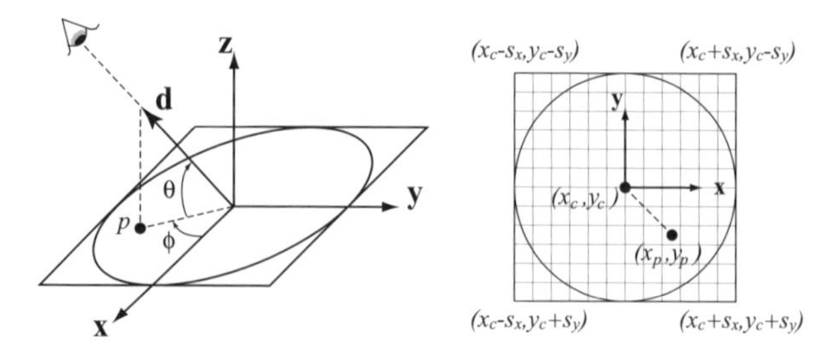

Figure 6. The process of converting viewing direction into the texture coordinates, or the XY-map projection.

calculate the texture coordinate (x,y) by the orthographic projection of **d** onto the plane defined by vectors **x** and **y**

$$x_p = s_x(\mathbf{d} \cdot \mathbf{x}) + x_c, \qquad y_p = s_y(\mathbf{d} \cdot \mathbf{y}) + y_c \,, \tag{9}$$

where the scale-and-bias parameters (s_x, s_y, x_c, y_c) represent the size and relative location of the view map on the texture map. This projection, as shown in Figure 6, is normally referred to as an XY-map. This projection offers a reasonable approximation on the interpolation quality and it is quite efficient to compute. Other transformations, such as the hyperbolical maps described in Heidrich *et al.* [1], can also be used for the viewing projection of light field maps.

Figure 7 illustrates the 3 light field maps pairs used to compute one approximation term of a light field for triangle Δ_t. The shaded portions indicate the parts of the light field maps that are used to decode appearance from a certain viewing direction. The middle column shows surface maps $g^{v_j}[r_p, s_p]_{\Delta t}$ and the right column shows view maps $h^{v_j}[\theta_q, \phi_q]$. The samples on the view maps are parameterized with the XY-map of the local reference frame attached to the vertex. Based on where the camera is located, the rendering algorithm calculates the texture coordinates $(x_i^{v_j}, y_i^{v_j})$ for each view map. To this end, we apply Equations (9) to the viewing direction vector $\mathbf{d}_i$ in the local reference frame $(\mathbf{x}_j, \mathbf{y}_j, \mathbf{z}_j)$ of vertex v_j to calculate the texture coordinate $(x_i^{v_j}, y_i^{v_j})$ on the XY-maps. This results in 3 texture fragments shown in the right column of Figure 7. Note that the texture coordinates are different for each view map fragment because we use different reference frames to compute them. The surface map texture coordinates do not depend on the viewing angle and they remain static throughout the rendering process.

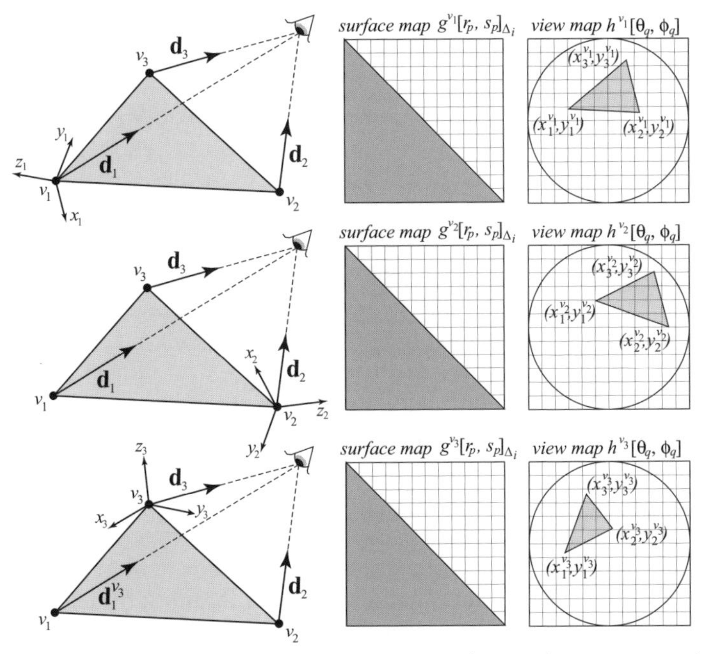

Figure 7. Calculation of light field maps texture coordinates for one approximation term of one triangle. Vertex reference frames are shown in the left column.

Evaluating one complete approximation term then simply proceeds as follows. We texture map each pair of surface map and view map texture fragments and multiply the results pixel-by-pixel. The product is then placed into the accumulation buffer. Multiple term approximation of each triangle light field is computed by running the same algorithm multiple times using their corresponding light field maps.

Utilizing Hardware Features for Efficient Rendering

The light field maps decoding process is simple and amendable for hardware implementation. In this section we discuss efficient rendering algorithms using specific hardware features, such as multitexturing, extended color range, and vertex shaders.

Multitexturing Support. This hardware support enables us to compute the modulation, or multiplication, of multiple texture fragments very efficiently in one rendering pass. Consequently, for the NMF-based approximations, which contain strictly positive light field maps, we need 3 rendering passes to render each approximation term with multitexturing graphics hardware that supports 2 texture sources. Each rendering pass decodes the approximation from one of the three vertices v_j in Equation 8.

Extended Color Range. For the PCA-based approximation, which in general produces light field maps that contain negative values, rendering can benefit from graphics hardware that permits a change to the limits of the color range from the traditional [0,1]. Without extended range support, we may need up to four rendering passes for each full-range modulation [2]. Recently more hardware platforms are supporting extended color range [41], but the output results are normally clamped to positive values. We may use this feature to evaluate full-range modulation in two rendering passes as follows. Let M be the result of modulation of two texture fragments A and B. Let M_+ and M_- be the clamped modulation of fragments (A,B) and $(-A,B)$ respectively. We can compute M by subtracting the outputs of the two modulations $M = M_+ - M_-$.

Improving Memory Bus Efficiency

In modern graphics systems, the amount of texture memory are generally less than the total system memory, and for larger light field maps, texture swapping between system and texture memory may be required. In order to reduce the swapping overhead, individual light field maps can be tiled or mosaicked together into larger texture maps, or texture atlases. The rendering routine goes through each texture atlas and reconstructs approximations associated with this texture. To ensure optimal texture swapping, we devised the tiling routine in two steps. First, surface geometry of the model is divided into several groups. Then, light field maps corresponding to the same group are tiled together to generate texture atlases.

Model Segmentation. When the texture memory in the graphics subsystem is abundant, one simply tiles all surface maps and view maps into two large texture atlases. In practice, however, not only may the size of texture memory not be sufficient, the maximum size of a texture map is also limited. Therefore, we may divide the surface geometry into several groups of triangles and collect the view maps and surface maps within each group into texture atlases. Our current implementation segments the model into multiple pieces by running a breadth-first search algorithm on surface triangles. Each search generates a connected group of triangles, and the search stops when all triangles connected to the root node are visited, or when the number of triangles or vertices exceeds a user-defined size.

Texture Atlas Generation. After model segmentation, surface maps and view maps can be tiled into texture atlases. To simplify the problem, our current implementation generates fixed size view maps for the whole model, and we only allow a predefined set of surface map sizes during the resampling process. Same-size light field maps from the same approximation term are then tiled together into one texture atlas. Since one triangle requires three surface maps per approximation term, these maps are tiled in the same texture.

Assume that the surface geometry is divided into p groups. We denote the view map atlas for term k, group i as V_k^i. Let $[S_k^{i1}, S_k^{i2}, ..., S_k^{iq_i}]$ be the list of surface map atlases in group i. When we render the scene in the texture atlas centric ordering, each view map and surface map is loaded only once, and is thus optimal in terms of texture swapping cost.

COMPRESSION OF LIGHT FIELD MAPS

Approximation through matrix factorization described in earlier can be thought of as a compression method that removes local redundancy in the vertex light field function. The compression ratio of this method is closely related to the size of the surface primitives used for partitioning. Currently, we choose the size of triangles empirically to obtain about two orders of magnitude compression ratio through approximation while maintaining high approximation quality without using many approximation terms.

Despite the efficiency of light field maps representation, they are still redundant. First, individual maps are similar to each other, suggesting global redundancy of the data. Second, some of the light field maps have very little information content and can be compressed further using a variety of existing image compression techniques. For optimal run-time performance, compressed light field maps need to be decompressed on-the-fly during rendering. In this section we discuss several techniques that satisfy these criteria. Figure 8 gives an overview of the different types of compression

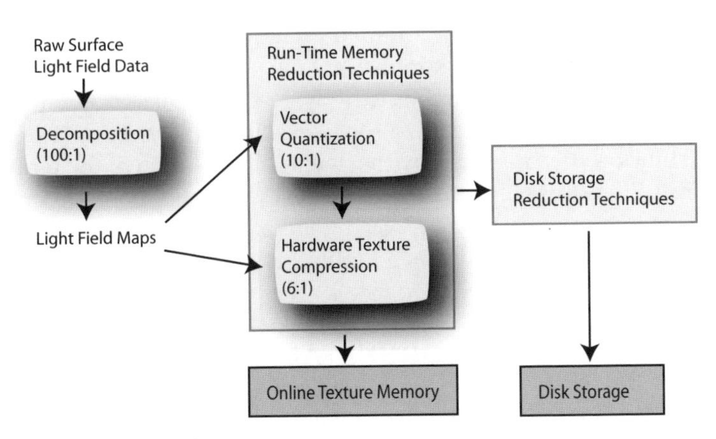

Figure 8. Compression Overview. The number under each technique describes its approximate compression ratio.

algorithms we have experimented with. Other image compression techniques can be used to further reduce the off-line storage size, but are not discussed here.

Global Data Redundancy. Data redundancy *across* individual light field maps can be reduced effectively using VQ [33]. Our implementation represents each triangle surface map $g^{v_j}_{\Delta_i}[r_p,s_p]$ and each view map $h^{v_j}[u_q,f_q]$ as a vector. The algorithm groups these vectors based on their size and generates a separate codebook for every group. We initialize the codebooks using either pairwise nearest neighbor or split algorithm. The codebooks are improved by the generalized Lloyd algorithm utilizing square Euclidean distance as the cost function. We then store the resulting codebooks as images. The rendering algorithm from VQ-compressed images does not change in any way-it simply indexes a different set of images.

We use either an user-specified compression ratio or the average distortion to drive the VQ compression algorithms. With the distortion-driven algorithm, the light field maps corresponding to the higher approximation terms exhibit more redundancy and thus are often compressed into a smaller codebook. In practice, light field maps can be compressed by an order of magnitude using VQ without significant loss of quality. In the current implementation, VQ is applied after all light field maps are computed. However, since the light field maps in the PCA algorithm are computed incrementally, we could potentially apply VQ after each iteration of the approximation algorithm and then factor the resulting error into the next approximation term.

Local Data Redundancy. Data redundancy *within* individual light field maps can be reduced efficiently using block-based algorithms. One such method, called S3TC™, is often supported on commodity graphics cards today. It offers compression ratios between 6:1 and 8:1 and can be cascaded with VQ for further size reduction. Limited by hardware implementation cost, these algorithms are not very sophisticated in nature. For example, the S3TC algorithm divides the textures into 4–by-4 texel blocks, and within each block it calculates and stores two representative colors. Each textel in the original block is then replaced by the linear interpolation of the representative colors. Since this algorithm uses blocks that are smaller than most light field maps, when compared to VQ, it generates noisier images but it preserves the specularities and sharp highlights better.

ACQUISITION OF SURFACE LIGHT FIELDS

In this Section, we describe the surface light fields acquisition systems used in our research. We first describe an accurate acquisition system for

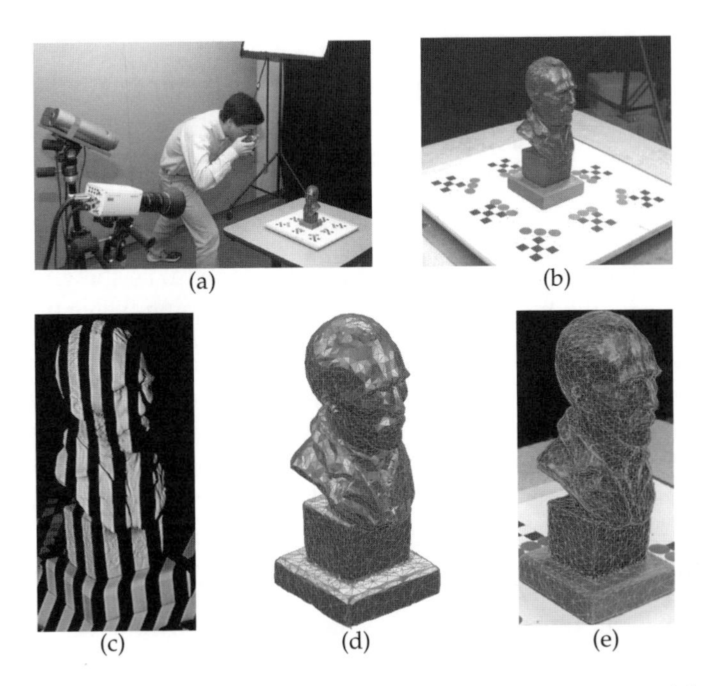

Figure 9. Small-scale scene acquisition. (a) The user is capturing images of the object under a fixed lighting condition using a hand-held digital camera. (b) One sample image. (c) The painted object being scanned using the structured lighting system. (d) The complete and simplified 3D triangular mesh constructed from 20 scans. (e) Reprojection of the triangular mesh onto image (b).

small-scale scenes, and move on to discuss another system designed to capture larger environments.

Small-Scale Acquisition

Figure 9 illustrates the small-scale acquisition system used in our research [10]. This system scans a 1 ft^3 volume accurately. It employs a registration platform for automatic registration between images and range scans. The object is placed on the platform and remains static to the platform throughout the acquisition process. The system scans geometry and radiance as two separate steps, and data from both steps are registered together using the coordinate system defined on the platform.

The first acquisition stage acquires radiance data with a hand-held camera, as shown in Figure 9(a). The internal parameters of the camera are calculated in a separate calibration stage. For each object, we capture between 200 to 400 images, covering the upper-hemisphere of the platform.

Figure 9(b) shows one sample image captured with this process. The color circles on the platform used to provides an initial estimate for the position of all the grid corners on the platform. The initial estimates are then localized using a corner finder to calculate the accurate camera pose relative to the platform. The outcome of this process is a set of N_I images captured from known vantage points in 3D space.

The 3D geometry of the object is scanned using a structured lighting system consisting of a projector and a video camera, as shown in Figure 9(a). In this phase, we paint the objects with white removable paint in order to improve the accuracy of the scanned geometry. Figure 9(c) shows an example camera image acquired during scanning. The projected stripped patterns observed by the camera are used to triangulate the 3D position of the object surfaces. Because each scan only covers parts of the object, we take between 10 and 20 around the object to completely cover the object surface. The individual scans are naturally registered together on the registration platform coordinate system. The resulting point cloud, containing approximately 500,000 points in this case, are fed into a mesh editing software [42] to reconstruct the final trinauglar surface mesh shown in Figure 9(d). Figure 9(e) shows the projection of the mesh onto the camera image displayed in Figure 9(b). The error of mesh reprojection is less than one out of two thousand pixels on the object silhouette.

Large-Scale Acquisition

The requirements for large-scale scene acquisition are very different from the small-scale ones. For example, the range scan system need to be capable of scanning a much larger volume. Also, it is often not desirable and practical to introduce calibration targets into the scene. Our large-scale surface light field acquisition system consists of a commercial laser rangefinder, the DeltaSphere™[43], for geometry acquisition, as shown in Figure 10(a), and a hand-held digital camera for radiance acquisition. The laser rangefinder acquires depth maps on a spherical coordinate system, and each scan produces approximately 8 million depth samples. Figure 9(c) shows one of the depth maps. Each range scan takes approximately 20 minutes. For this scene, we took a total of 7 panoramic depth scans and approximately 100 pictures at different positions in the environment. Figure 10(b) shows one example photograph after removing nonlinear distortions.

We use a commercial software package called Polyworks™to bring all scans into the same reference frame [44]. Polyworks implements an Iterative Closest Point (ICP) algorithm for point-cloud registration. After depth maps are registered, we triangulate each depth map separately and merge individual triangular mesh into a unified geometry with Polyworks.

(c) (d) (e)

Figure 10. Large-scale scene acquisition. (a) DeltaSphere rangefinder. (b) A picture of the environment after undistortion. (c) A depth map acquired by the rangefinder. Darker samples represent smaller depths. Low-confidence depth samples are colored in red for illustrative purposes.

We recover the camera poses using manual 2D-3D correspondences. For each image, a user first chooses a depth map, and then selects 6 or more pairs of corresponding points on both the depth map and the image. This allows the calculation of the external parameters with respect to the coordinate system of the depth map. Then, the transformations between individual depth maps and the global coordinate system obtained through the ICP algorithm during the geometry reconstruction stage is used to recover the global camera pose of the image. This process relies on an accurate ICP registration, which also requires accurate rangefinder calibration. Currently inaccuracies in rangefinder calibration account for most of the overall error. However, as shown in Figure 19, artifacts due to inaccurate registration can largely be resolved by the view-dependent nature of surface light fields representation.

RADIANCE DATA RESAMPLING

The approximation algorithms proposed earlier take densely sampled surface light field functions as input. In practice, however, data from the acquisition systems are scattered samples of the actual surface light field functions. Now we discuss methods and issues in preparing these data for the approximation algorithms. In particular, we assume that the input data consist of a triangular geometric mesh together with a set of camera images registered to the mesh. Alternatively, we may bypass the resampling process altogether at a cost of lower approximation quality, as described in [39].

The input data at this stage consist of a triangular mesh and a set of images taken with known camera internal and external parameters. The goal of resampling is to construct a surface light field function $f[r_p,s_p,\theta_q,\phi_q]$ that best represents the input data. The problem of surface light field resampling is, in general, a 4D data reconstruction problem. However, if the reprojection sizes of the triangles are relatively small compared to their distances to the camera, for a triangle the samples from the same image can be regarded as having identical viewing directions (θ, ϕ). Under this assumption, the resampling process can be approximated by a two-stage algorithm that first resamples on $[r_p,s_p]$ and then on $[\theta_q, \phi_q]$. We refer to the first stage as the *surface normalization* stage and the second stage as the *view interpolation* stage. We focus our discussion on one vertex light field function $f^{v_j}[r_p,s_p,\theta_q,\phi_q]$.

Surface Normalization. Before normalization, we need to determine the visible cameras for the target vertex v_j. Repeating the visibility calculation for all N camera images yields a list of N_j visible vertex views. We denote the viewing directions of the visible views as $[u_{vj}, f_{vj}]$, where the index v^j = $1,...N_j$. The visible vertex views correspond to a set of texture patches of irregular size captured from various viewing directions. The algorithm then normalizes each texture patch to have the same shape and size as the others by using bilinear interpolation of the pixels in the original views. In order to preserve image sampling rate, the size of the normalized patch is chosen to be proportional to the size of the largest projected view. Then, each vertex view is multiplied with the hat function L^{vj} described in equation 3.

View Interpolation. At this stage, each triangle view contains a uniform number of samples, but the sampling of views is still irregular. We denote the input function at this stage as $t^{v_j}[r_p, s_p, u_{vj}, f_{vj}]$. Given the input function, the goal is to reconstruct the vertex light field function $f^{v_j}[r_p, s_p, u_q, f_q]$. To do this, we construct an interpolation function $\mathbf{Q}^j(u,f)$ that returns a 3×2 matrix whose first and second columns define the indices to the original views and the interpolation weights respectively. The components of $\mathbf{Q}^j(u,f)$ have the following properties

$$\forall \{\theta, \phi \mid -\pi \leq \theta \leq \pi, -0.5\pi \leq \phi \leq 0.5\pi\},$$

$$\begin{cases} \mathbf{Q}^j(\theta,\phi)_{k1} & \in \quad \{1,2,\ldots N_j\}, \\ \displaystyle\sum_{k=1}^{3} \mathbf{Q}^j(\theta,\phi)_{k2} & = \quad 1. \end{cases} \qquad (10)$$

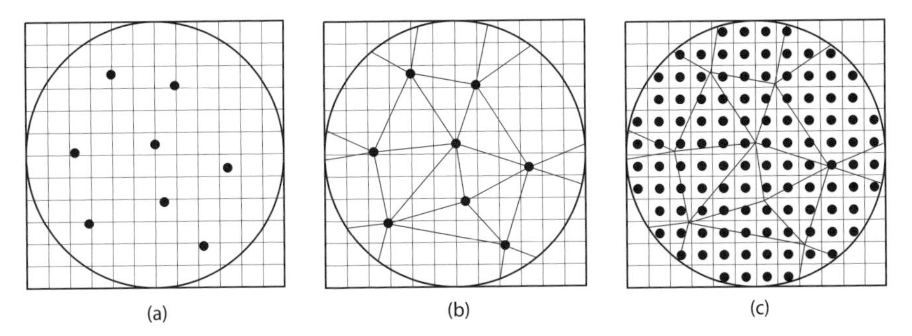

Figure 11. View Interpolation Process. (a) Projection of original views, (b) Delaunay triangulation of projected views, (c) uniform grid of views computed by blending the original set of views.

This allows us to perform view interpolation as follows

$$\begin{bmatrix} i_1 & w_1 \\ i_2 & w_2 \\ i_3 & w_3 \end{bmatrix} \leftarrow \mathbf{Q}^j(\theta_q, \phi_q)$$
$$f^{v_j}[r_p, s_p, \theta_q, \phi_q] = \sum_{k=1}^{3} w_k \tau^{v_j}[r_p, s_p, \theta^j_{i_k}, \phi^j_{i_k}]. \tag{11}$$

Figure 11 describes the the definition of the interpolation function $\mathbf{Q}_j$. First, the viewing directions from the visible views $[\theta_{vj}, \phi_{vj}]$ are projected onto the xy plane using XY-map projection. The result is a set of texture coordinates, as shown in Figure 11(a). These coordinates are used to generate a Delaunay triangulation as shown in Figure 11(b). We then define the interpolation function as follows. For each point on the XY-map, the indices i_k are the three vertices of the triangle surrounding it, and the weights w_k are the barycentric coordinates of this point within this triangle. The resampled directions $[\theta_p, \phi_p]$ are the regular grid shown in Figure 11(c). If we collapse 2D indices $[r_p, s_p]$, $[\theta_q, \phi_q]$, and $[\theta_{vj}, \phi_{vj}]$ into one dimension respectively similar to Equation 5, we can rewrite Equation 11 using 2D matrices as follows

$$\mathbf{F}^{v_j} = \mathbf{I}^{v_j}\mathbf{W}^{v_j}. \tag{12}$$

Since each column of matrix $\mathbf{W}^{v_j}$ contains at most 3 non-zero rows, the above computation can be computed efficiently using sparse matrix multiplication routines.

RESULTS

We have acquired objects with diverse and complex reflection properties. The Van Gogh *bust* shown in Figure 17 is approximately one foot tall. The simplified surface geometry does not contain details such as chisel marks, but our experiments show that these details are modelled quite well by using images only. The *dancer* shown in Figure 16 has a metallic look except on the blouse and the skirt of the model. The topology of this object introduces interesting effects such as self shadowing. The *star* shown in Figure 18 is approximately 1/2 feet tall and made out of glass covered with twirled engravings and thin layers of paint. Depending on the viewing angle, it is either semi-transparent or anisotropically reflective. Such reflectance properties are difficult to model analytically. The toy *turtle* shown in Figure 15 is covered with a velvet-like material that is normally difficult to represent using traditional techniques.

Light Field Mapping is inherently scalable to larger scenes, because each vertex light field function and process each function independently. Figure 1 shows a synthetic surface light field composed with scanned objects. Images from different view points are rendered using a commercial renderer 3D Studio Max™. These images are then treated as input photographs for the LFM data processing pipeline. The resulting scene can be rendered at interactive rate, or at approximately one thousand times faster than the commercial renderer used to generate input images. Figure 19 shows the 3–term PCA approximation of an office scene scanned using our large-scale acquisition system. This scene can also be rendered at interactive rates on a PC.

PARTITIONING AND APPROXIMATION

We experimented with both vertex-centered and triangle-centered partitioning methods, as illustrated in Figure 12. In this experiment, we use rendered synthetic images to exclude artifacts caused by acquisition error. Clearly, triangle-centered approximations produce discontinuity artifacts across triangle boundaries. After adding more approximation terms, such artifact becomes less obvious but still visible. This problem cannot be corrected by adding more approximation terms. On the other hand, vertex-centered partitioning does not exhibit similar artifact even when only one approximation term is used.

To measure the quality of surface light field approximations, we calculate the RMS difference between the original and approximated light field matrices. We also calculate Peak Signal-To-Noise Ratio (PSNR), a commonly used image quality metric directly related to RMS error by

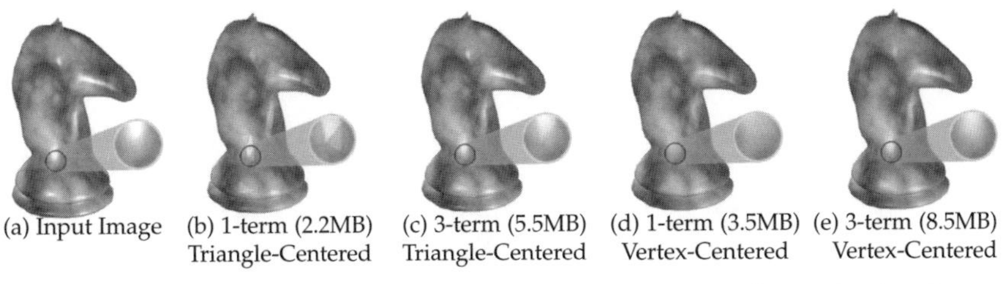

Figure 12. Comparison between triangle- and vertex-centered partitioning. Input picture (a) is a synthetic chess piece rendered using 3D Studio Max™. A total of 256 input pictures are used to compute the PCA-based light field maps used to render (b)-(e).

$$\mathrm{PSNR} = 20 \log_{10} \left(\frac{I_{max}}{\mathrm{RMS}} \right), \tag{13}$$

where I_{max} represents the maximum pixel intensity. Figure 13 shows the approximation quality for both PCA and NMF algorithms. For each object, a PSNR is calculated over *all* of its vertex light field matrices. In this figure, we use 24–bit RGB pixels and therefore $I_{max} = 2^8-1 = 255$. As shown in the figure, both techniques provide high quality approximations using very small number of terms. Between the two algorithms, PCA produces better quality than NMF. However, the difference is visually almost indistinguishable.

Rendering

Figure 14 compares the rendering performance of PCA-based and NMF-based approximations. On this platform, rendering a full-range

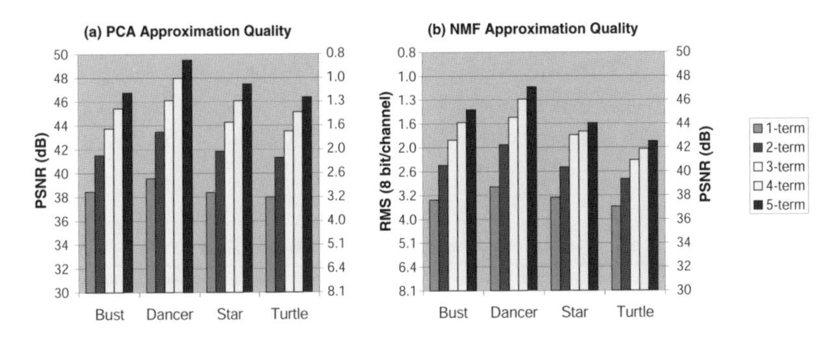

Figure 13. Approximation quality for different models and different number of decomposition terms. PSNR and RMS are based on the weighted average of the approximation errors for all light field matrices.

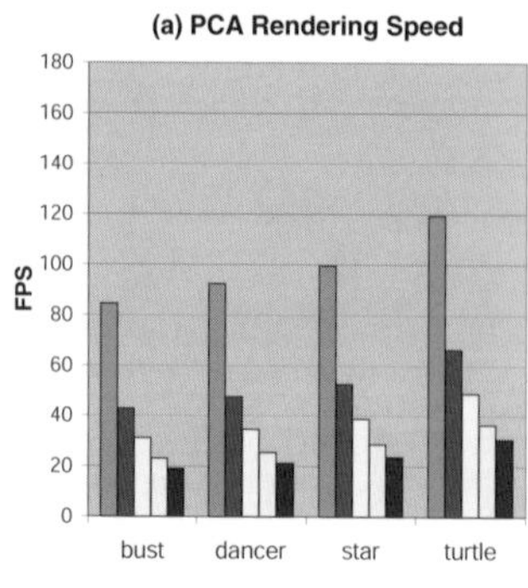

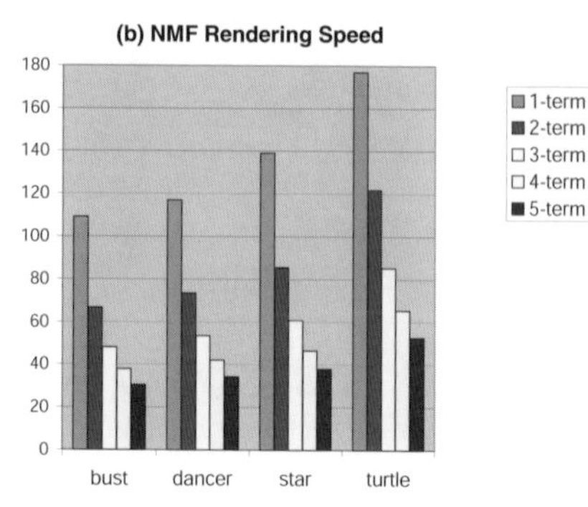

Figure 14. Rendering performance using NVIDIA GeForce 3 graphics card on a 2GHz Pentium 4 PC displayed at 1024×768 window with objects occupying approximately 1/3 of the window.

multitexturing modulation requires 2 rendering passes. Excluding the diffuse layer, we need 6*K* rendering passes for a *K-term* PCA approximation, whereas only 3K passes is required for a *K-term* NMF approximation. We observe that NMF-based rendering is 50% faster than PCA-based for the same number of approximation terms. The performance disparity would be larger if the target platform supported only positive texture values. The rendering performance is not very sensitive to the size of light field map data-doubling the image size reduces the frame rate by less than 20%. Rendering from compressed and uncompressed light field maps are equally fast if image sets in both cases fit into the texture cache.

Compression

Table 1 provides information on the data sizes of the models used in the experiments. We use 24–bit RGB images in all the experiments. The input image size in this table represents the total size of input images from the acquisition system. The effective image size represents the size of all *foreground* pixels of the input images. The resampled data size represents the size of actual surface light field function used for approximation. In this

Table 1. The sizes of the experimental data sets.

Models	Triangles	Number of Images	Input Image Size	Effective Image Size	Resampled Data Size	Sampling Density (θ, ϕ)
Bust	6531	339	2.5GB	289 MB	5.1 GB	32×32
Dancer	6093	370	2.7 GB	213 MB	2.9 GB	32×32
Star	4960	282	2.1 GB	268 MB	5.2 GB	32×32
Turtle	3830	277	1.7 GB	233 MB	3.8 GB	32×32

Table 2. The size and compression ratio of the radiance data obtained through the light field map approximation and additional compression of the surface light field maps.

Models	Light Field Maps (3–term)	Compression of Light Field Maps VQ	S3TC	VQ+S3TC
Bust	47.7 MB (106:1)	5.7 MB (885:1)	8.0 MB (636:1)	951 KB (5475:1)
Dancer	35.7 MB (82:1)	5.4 MB (542:1)	5.9 MB (490:0)	904 KB (3303:1)
Star	42.3 MB (122:1)	7.2 MB (716:1)	7.0 MB (737:1)	1.2 MB (4301:1)
Turtle	31.7 MB (121:1)	4.5 MB (847:1)	5.3 MB (726:1)	755 KB (5084:1)

process, viewing directions are resampled on a 32×32 pixel grid. The resulting resampled data size is approximately twice as large as the input images. Traditionally, research on light field compression reports results based on the size of the resampled light field function [5,34], and therefore we calculate the compression ratio also based on the resampled data.

Table 2 lists the size and compression ratio of the light field data obtained through light field maps approximation and additional compressions of the light field maps. For all the objects, the size of the geometric data falls below 10KB when compressed using topological surgery [45] and therefore is negligible compared to the size of light field maps. By combining VQ with S3TC hardware texture compression, our method achieves a *run-time* compression ratio of over 5000:1 for a 3–term approximation. For interactive purposes, 1–term approximation is often sufficient and thus the resulting compression ratio approaches 4 orders of magnitude.

Figures 15–18 compare the rendering quality of our routines against the input images and report the corresponding image errors. The errors re-

Figure 15. Comparison for the *turtle* model between the input images shown at the top row and the images synthesized from the 1–term PCA approximation compressed using both VQ and S3TC shown at the bottom row. APE = 9.5, PSNR = 25.5 dB, the compression ration is 8202:1, and the size of compressed light field maps is 468 KB.

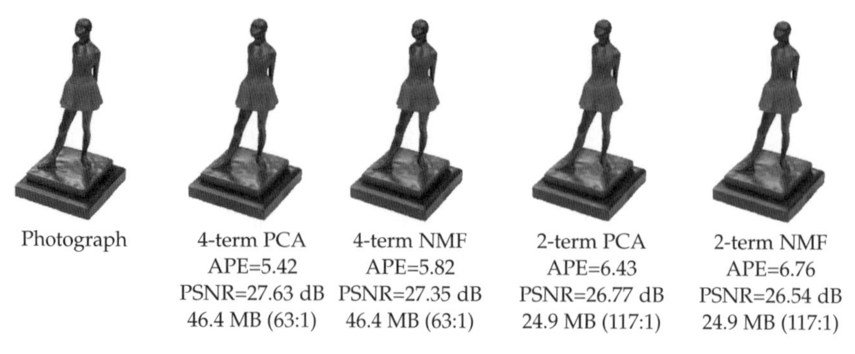

Figure 16. Comparison between PCA and NMF approximation methods. Using the same number of terms, PCA light field maps produce less error, but are slower to render than NMF.

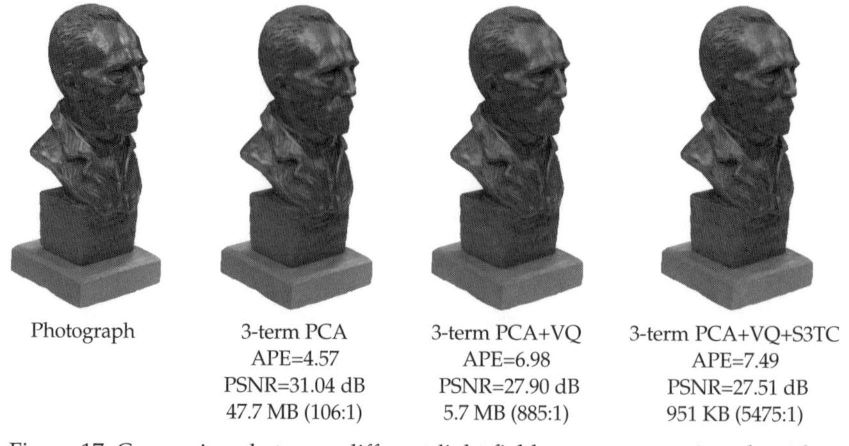

Figure 17. Comparison between different light field map compression algorithms using the *bust* model. VQ tends to diminish the highlight while S3TC preserves highlights better at the expense of color quality.

ported in the figures are computed based on the differences between the input images and the rendered images using both Average Pixel Error (APE) and PSNR for the *foreground* pixels only. The image errors are larger than the approximation error in Figure 13 due to the resampling process. In this process, if samples from the input images do not coincide with the viewing parameter grid, the original samples is lost and we can not reconstruct input images from the resampled surface light field functions.

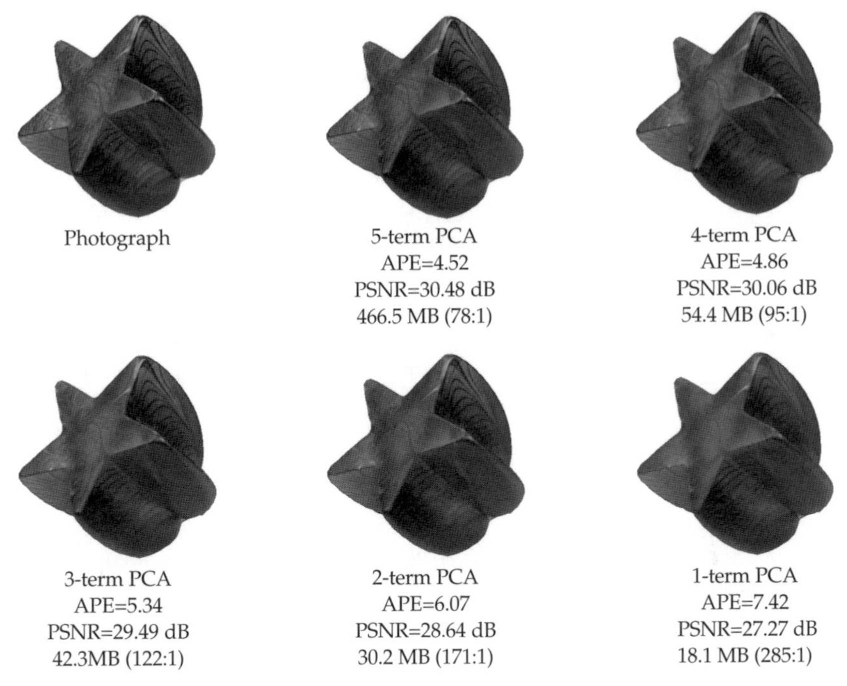

Figure 18. The figure demonstrates the progressive nature of PCA approximation. The same *star* model is rendered using different number of approximation terms.

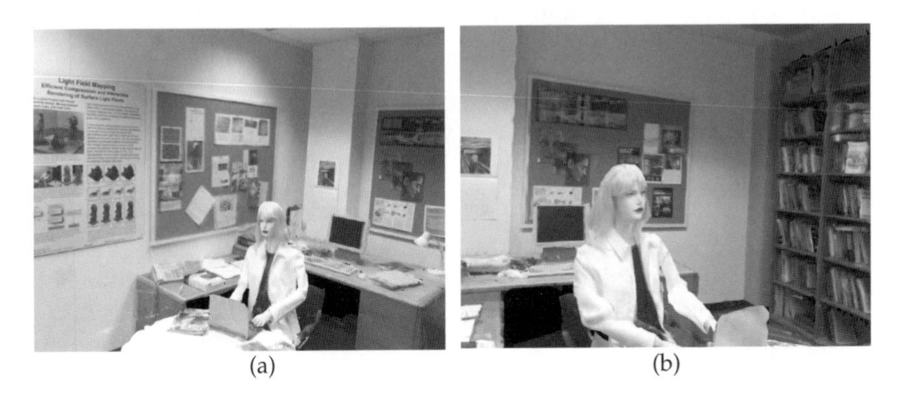

Figure 19. A real office scene acquired using our large-scale acquisition system and rendered with 3–term PCA light field maps approximation. The physical dimensions of the office are approximately 15*ft(W)*×10*ft(D)*×8*ft(H)*.

CONCLUSIONS AND FUTURE WORK

We have developed a new representation of surface light fields and demonstrated its effectiveness on both synthetic and real data. Using our approach, surface light fields can be compressed several thousand times and efficiently rendered at interactive speed on modern graphics hardware directly from their compressed representation. Simplicity and compactness of the resulting representation leads to a straightforward and fully hardware-accelerated rendering algorithm. Additionally, we present a new type of light field data factorization that produces positive only factors. This method allows faster rendering using commodity graphics hardware. Furthermore, this article contains a detailed explanation of the data acquisition and preprocessing steps, providing a description of the complete modeling and rendering pipeline. Finally, our PCA-based approximation technique is particularly useful for network transport and interactive visualization of 3D photography data because it naturally implies progressive transmission of radiance data.

Our current implementations of light field maps compression algorithms only demonstrated part of a variety of possible algorithms, and this issue certainly deserves more attention. One of the limitations of the surface light field is that it parameterizes only the outgoing radiance of the data. We are planning to work on extending our work to relighting and animation of image-based models. If successful, these results would prove that image-based modeling and rendering is a practical and an appealing paradigm for enhancing the photorealism of interactive 3D graphics.

ACKNOWLEDGEMENTS

We thank Srihari Sukumaran for helping with the early version of the resampling and tiling algorithm. We are indebted to Omid Moghadam for his support in making light field mapping part of the MPEG4 standard. We thank Henry Fuchs, Anselmo Lastra, Ara Nefian, Lars Nyland and Herman Towles for helpful suggestions and discussions and Gary Bradski, Bob Liang and Justin Rattner for encouraging this work. A significant portion of this article appeared in [10] and we thank Radek Grzeszczuk and Jean-Yves Bouguet for their significant contributions.

This research is in part supported by NSF Cooperative Agreement "Science and Technology Center for Computer Graphics and Scientific Visualization", "National Teleimmersion Initiative" funded by Advanced Networks and Services, and a Link Foundation Fellowship.

REFERENCES

[1] W. Heidrich and H.-P. Seidel, "Realistic, Hardware-Accelerated Shading and Lighting," in *Proceedings of ACM SIGGRAPH 1999,* pp. 171–178, August (1999).

[2] J. Kautz and M. D. McCool, "Interactive Rendering with Arbitrary BRDFs using Separable Approximations," *Eurographics Rendering Workshop 1999,* June (1999).

[3] J. Kautz and H.-P. Seidel, "Towards Interactive Bump Mapping with Anisotropic Shift-Variant BRDFs," *2000 SIGGRAPH / Eurographics Workshop on Graphics Hardware,* pp. 51–58, August (2000).

[4] L. McMillan and G. Bishop, "Plenoptic Modeling: An Image-Based Rendering System," in *Proceedings of ACM SIGGRAPH 1995,* pp. 39–46, August (1995).

[5] M. Levoy and P. Hanrahan, "Light Field Rendering," in *Proceedings of ACM SIGGRAPH 1996,* pp. 31–42, August (1996).

[6] S. J. Gortler, R. Grzeszczuk, R. Szeliski, and M. F. Cohen, "The Lumigraph," in *Proceedings of ACM SIGGRAPH 1996,* pp. 43–54, August (1996).

[7] C. Buehler, M. Bosse, L. McMillan, S. Gortler, and M. Cohen, "Unstructured Lumigraph Rendering," in *Proceedings of ACM SIGGRAPH 2001,* pp. 425–432, August (2001).

[8] G. S. P. Miller, S. Rubin, and D. Ponceleon, "Lazy Decompression of Surface Light Fields for Precomputed Global Illumination," *Eurographics Rendering Workshop 1998,* pp. 281–292, June (1998).

[9] D. N. Wood, D. I. Azuma, K. Aldinger, B. Curless, T. Duchamp, D. H. Salesin, and W. Stuetzle, "Surface Light Fields for 3D Photography," in *Proceedings of ACM SIGGRAPH 2000,* pp. 287–296, July (2000).

[10] W.-C. Chen, J.-Y. Bouguet, M. Chu, and R. Grzeszczuk, "Light Field Mapping: Efficient Representation and Hardware Rendering of Surface Light Fields," in *Proceedings of ACM SIGGRAPH 2002,* July (2002).

[11] G. J. Ward, "Measuring and Modeling Anisotropic Reflection," in *Proceedings of ACM SIGGRAPH 1992,* pp. 265–272, July (1992).

[12] P. Schröder and W. Sweldens, "Spherical Wavelets: Efficiently Representing Functions on the Sphere," in *Proceedings of ACM SIGGRAPH 1995,* pp. 161–172, August (1995).

[13] J. Koenderink and A. van Doorn, "Bidirectional Reflection Distribution Function Expressed in Terms of Surface Scattering Modes," in *European Conference on Computer Vision,* pp. II:28–39, (1996).

[14] E. P. F. Lafortune, S.-C. Foo, K. E. Torrance, and D. P. Greenberg, "Non-Linear Approximation of Reflectance Functions," in *Proceedings of ACM SIGGRAPH 1997,* pp. 117–126, August (1997).

[15] A. Fournier, "Separating Reflection Functions for Linear Radiosity," *Eurographics Rendering Workshop 1995,* pp. 296–305, June (1995).

[16] K. E. Torrance and E. M. Sparrow, "Polarization, Directional Distribution, and Off-Specular Peak Phenomena in Light Reflected from Roughened Surfaces," *Journal of Optical Society of America,* vol. 56, no. 7, (1966).

[17] X. D. He, K. E. Torrance, F. X. Sillion, and D. P. Greenberg, "A Comprehensive Physical Model for Light Reflection," in *Proceedings of ACM SIGGRAPH 1991* (T. W. Sederberg, ed.), pp. 175–186, July (1991).

[18] R. L. Cook and K. E. Torrance, "A Reflectance Model for Computer Graphics," *ACM Transactions on Graphics,* vol. 1, pp. 7–24, January (1982).

[19] P. Poulin and A. Fournier, "A Model for Anisotropic Reflection," in *Proceedings of ACM SIGGRAPH 1990,* pp. 273–282, August (1990).

[20] M. D. McCool, J. Ang, and A. Ahmad, "Homomorphic Factorization of BRDFs for High-Performance Rendering," in *Proceedings of ACM SIGGRAPH 2001* (E. Fiume, ed.), pp. 171–178, ACM Press / ACM SIGGRAPH, August (2001).

[21] D. D. Lee and H. S. Seung, "Learning the Parts of Objects by Non-Negative Matrix Factorization," *Nature,* vol. 401, pp. 788–791, (1999).

[22] K. J. Dana, B. van Ginneken, S. K. Nayar, and J. J. Koenderink, "Reflectance and texture of real-world surfaces," *ACM Transactions on Graphics,* vol. 18, pp. 1–34, Jan. (1999).

[23] X. Liu, H.-Y. Shum, and Y. Yu, "Synthesizing Bidirectional Texture Functions for Real-World Surfaces," in *Proceedings of ACM SIGGRAPH 2001,* pp. 97–106, Aug. 12–17 (2001).

[24] D. K. McAllister, *A Generalized Surface Appearance Representation for Computer Graphics.* PhD thesis, Department of Computer Science, University of North Carolina—Chapel Hill, May (2002).

[25] H. P. A. Lensch, J. Kautz, M. Goesele, W. Heidrich, and H.-P. Seidel, "Image-Based Reconstruction of Spatially Varying Materials," in *Twelveth Eurographics Rendering Workshop 2001,* pp. 104–115, Eurographics, June (2001).

[26] L. McMillan, *An Image-Based Approach to Three-Dimensional Computer Graphics.* PhD thesis, Department of Computer Science, University of North Carolina—Chapel Hill, May (1997).

[27] W. R. Mark, L. McMillan, and G. Bishop, "Post-Rendering 3D Warping," in *1997 Symposium on Interactive 3D Graphics,* pp. 7–16, Apr. (1997).

[28] C.-F. Chang, G. Bishop, and A. Lastra, "LDI Tree: A Hierarchical Representation for Image-Based Rendering," in *Proceedings of ACM SIGGRAPH 1999,* pp. 291–298, August (1999).

[29] K. Pulli, M. Cohen, T. Duchamp, H. Hoppe, L. Shapiro, and W. Stuetzle, "View-based Rendering: Visualizing Real Objects from Scanned Range and Color Data," *Eurographics Rendering Workshop 1997,* pp. 23–34, June (1997).

[30] P. E. Debevec, Y. Yu, and G. D. Borshukov, "Efficient View-Dependent Image-Based Rendering with Projective Texture-Mapping," *Eurographics Rendering Workshop 1998,* pp. 105–116, June (1998).

[31] B. Heigl, R. Koch, M. Pollefeys, J. Denzler, and L. V. Gool, "Plenoptic Modeling and Rendering from Image Sequences Taken by a Hand-Held Camera," in *DAGM99,* pp. 94–101, (1999).

[32] J.-X. Chai, X. Tong, S.-C. Chan, and H.-Y. Shum, "Plenoptic Sampling," in *Proceedings of ACM SIGGRAPH 2000,* pp. 307–318, July (2000).

[33] A. Gersho and R. M. Gray, *Vector Quantization and Signal Compression.* Kluwer Academic Publishers, (1992).

[34] M. Magnor and B. Girod, "Data Compression for Light Field Rendering," *IEEE Trans. Circuits and Systems for Video Technology,* vol. 10, pp. 338–343, April (2000).

[35] P. Eisert, T. Wiegand, and B. Girod, "Model-Aided Coding: A New Approach to Incorporate Facial Animation into Motion-Compensated Video Coding,"

IEEE Trans. Circuits and Systems for Video Technology, vol. 10, pp. 344–358, April (2000).

[36] B. Girod, "Two Approaches to Incorporate Approximate Geometry Into Multi-view Image Coding," in *International Conference on Image Processing,* p. TA01.02, (2000).

[37] K. Nishino, Y. Sato, and K. Ikeuchi, "Appearance Compression and Synthesis based on 3D Model for Mixed Reality," in *International Conference on Computer Vision,* pp. 38–45, (1999).

[38] C. M. Bishop, *Neural Networks for Pattern Recognition.* Clarendon Press, (1995).

[39] W.-C. Chen, *Light Field Mapping: Efficient Representation of Surface Light Fields.* PhD thesis, Department of Computer Science, University of North Carolina—Chapel Hill, June (2002).

[40] G. H. Golub and C. F. V. Loan, *Matrix Computations.* Johns Hopkins University Press, 3rd ed., (1996).

[41] J. Spitzer, "Texture Compositing With Register Combiners," in *Game Developers Conference,* April (2000).

[42] Raindrop, "Raindrop Geomagic Inc. Geomagic Studio 3.0.," http://www.geomagic.com/products/studio/. (1999).

[43] 3rdTech, "3rdTech Inc. DeltaSphere 3D Scene Digitizer," http://www.3rdtech.com/DeltaSphere.htm. (2000).

[44] InnovMetrics, "InnovMetrics Inc. Polyworks 6.0," http://www.innovmetric.com/. (2001)).

[45] G. Taubin and J. Rossignac, "Geometric Compression Through Topological Surgery," *ACM Transactions on Graphics,* vol. 17, pp. 84–115, April (1998).

Real-Time Interaction with Spline Surfaces

Jatin Chhugani

The Johns Hopkins University

224 New Engineering Building

3400 N. Charles St.

Baltimore, MD 21218

Research Advisor: Dr. Subodh Kumar

ABSTRACT

We present algorithms for real time interaction with models consisting of tens of thousands of spline surfaces. To exploit the fast triangle rendering capability of the modern hardware, the algorithm performs a dynamic adaptive triangulation of spline surfaces. The algorithm is dynamic as it adjusts the level of detail of the triangulation every frame, and is adaptive as it samples region of high curvature more densely as compared to the regions with lower curvature. Additionally, we also investigate the use of points as primitives used to display curved surfaces. This algorithm samples the spline surfaces and directly renders the sample points at varying point sizes to guarantee a desired frame rate or a desired accuracy. Another aspect to interactive rendering is to conservatively discard invisible geometry early in the pipeline and to remove unnecessary detail not visible from the current viewpoint. We propose an algorithm, which combines the advantages of both these methods to provide further speedups in the rendering time of large geometric models.

INTRODUCTION

Interactive rendering of surface models is an important area of computer graphics research. Interaction with a simulation environment involves not only rendering a complex scene at a high frame rate, but also computing proximity (e.g. collision detection, intersection, minimum separation distance, etc.) between the user and the objects (in case of simulated walkthroughs) or amongst the objects themselves for dynamic scenes. Curved surface models are used in applications ranging from CAD industry and medical and scientific visualization to entertainment industry including computer games like Quake-III [1] and Max Payne [2], for a variety of simulations. Splines, especially of the Non-Uniform Rational B-Spline (NURBS) and Bezier forms [3], remain the representation of choice for a large class of these applications. For example, submarines, airplanes, automobiles, etc. are commonly modeled as splines. Hence, spline rendering has been an active area of research for over two decades. Ray tracing [4–6], pixel level subdivision [7, 8], and scan-line based algorithms [9–11] have been used in the past to render surfaces. However, none of these are quite efficient on the current graphics systems. In order to exploit fast triangle rendering hardware, research in the last decade has focused on generating polygonal approximations of surfaces using uniform tessellation [12–14], or adaptive tessellation [15, 16]. The adaptive tessellation algorithms are intended for off-line static triangulation of models. A one-time static tessellation (in sufficient detail) of many real-world models, would require hundreds of million of triangles [17]. Schemes based on view-dependent mesh simplification [18, 19] are commonly used to improve rendering speed on any given graphics system. Typically, these require management of large amounts of data and are still unable to generate more detail than the initial tessellation. Note that these methods first generate a large number of triangles and subsequently reduce the count. We call these 'backward' techniques. 'Forward techniques', on the other hand, perform view-dependent tessellation [12], with the advantage being the ability to get arbitrarily high precision. However, considerable time must be allocated to evaluate the new samples at each frame and to update the triangulation. As a result, only simple surface sampling algorithms, *e.g.* uniform domain sampling, are used, thus causing over-tessellation for many areas of the model. The backward techniques, on the other hand, are able to utilize significant resources in a pre-processing step to limit dense sampling to areas of high curvature.

However, experience shows that in order to guarantee a small screen-space error, we are forced to use triangles that are, on average, small in screen space. Moreover, small triangles are typically more efficiently rendered as points. Recently, the possibility of rendering points has been gain-ing popularity. The method is especially well suited for models acquired through 3D scanning [20]. The idea of using points as display primitives for continuous surfaces was introduced in 1985 by M. Levoy and T. Whitted [21], and more recently has been explored further in [22–24]. These techniques use hierarchical data structures and forward warping to store and render the point data efficiently. Wand et. al. [25] present an output sensitive rendering algorithm, which takes as input a complex triangular model and renders dynamically chosen set of surface sample points. Also algorithms that mix triangle and point rendering have been proposed [26]. Our goal is to efficiently and accurately display spline models. The tessellation-based schemes have accurate projections only at the vertexes of the triangles. One way to compromise is to tessellate the surface into nominally curved 'elements', which follow the surface more closely than a triangle may. Unfortunately, it is difficult to match up complex elements at their common boundaries. One solution is to let them overlap as long as they are close enough to each other in the region of overlap. We have taken this approach. We propose the first algorithm for rendering spline surfaces using such element tessellation. We have been able to factor out most of the computation to the pre-processing stage. At rendering-time, we just need to choose from a set of pre-computed samples and determine their element sizes.

Another advantage of tessellating parametric surfaces using triangles and points lies in the advantages these representations offer in solving the proximity queries in real-time, within a specified error. Computing minimum distance between two parametric surfaces or computing collision between such surfaces involves solving high degree equations, which may take a lot of time. However, by tessellating these models, one can query these coarse representations of the complex surfaces to get the solution. In case of stringent thresholds, one can further refine the tessellation to compute the solutions. Hence tessellating offers benefits not only in quick rendering but also for faster proximity analysis.

Geometric modeling of real world data has made significant advances in recent years. Increasingly complex models are being generated that need to be visualized interactively for various simulations. While the triangle rendering performance of the state of the art graphics hardware has been improving, model sizes have grown much faster to tens, and even hundreds, of million triangles. We demonstrate that arbitrarily high hardware rendering performance is not necessary to display models with high complexity. The basic premise of this argument is that display devices have limited resolution, and that limits the detail to which the model can be displayed. This implies, that in practice, if the correct set of triangles were displayed, a small number of them would suffice. An implication of this result is that the right algorithm can accurately display arbitrarily sized models on a 640 × 480 pixel screen using a commodity graphics card that can display 10 million polygons per second. Of course, this algorithm would have to display only the visible objects at the right level of detail, both areas of active research. Aliaga et. al. [17] present MMR, the first comprehensive rendering system that combines different methods like Level of Detail computation, visibility and image impostors. We present a framework that derives the complete advantage of both simplification and visibility computation at the same time. While our implementation makes a specific choice of simplification and visibility, in fact many different geometric error-based simplification techniques and cell-based visibility algorithms can be plugged into our framework. We partition the space into view-cells and, for each cell, pre-compute and store its vLOD: the triangles visible from the cell at the appropriate detail. We show that neither the preprocessing time, nor the disk storage requirement is prohibitive.

Mathematical Background

A rational tensor-product Bezier patch, $\mathbf{F}(u,v)$ of degree $m \times n$ is defined for $(u,v) \in [0,1] \times [0,1]$, and specified by a mesh of control points, $\mathbf{p}_{ij}$ and their weights, w_{ij}, $0 \le i \le m, 0 \le j \le n$:

$$F(u,v) = \frac{\sum_{i=0}^{m}\sum_{j=0}^{n} w_{ij} p_{ij} B_i^m(u) B_j^n(v)}{\sum_{i=0}^{m}\sum_{j=0}^{n} w_{ij} B_i^m(u) B_j^n(v)}, \qquad (1)$$

where the Bernstein function $B^{mi}(t) = {}^{m}C_i t^i (1-t)^{m-i}$.

The normal direction is given by $\mathbf{N}(u,v) = \mathbf{F}_u \times \mathbf{F}_v$, where $\mathbf{F}_u$ and $\mathbf{F}_v$ are the partial derivatives in the u and v directions respectively. The Delaunay triangulation of points on a plane has the property that the circumcircle of no triangle contains any other point of the set of vertexes being triangulated. This property clearly avoids long and skinny triangles. In fact, Delaunay triangulation produces maximum possible smallest-internal angle of any triangle.

Organization

In the rest of the paper, we assume familiarity with NURBS and Bezier surfaces and Delaunay triangulation. We start by explaining our view-dependent adaptive tessellation algorithm. This is followed by our approach to render parametric surfaces using point rendering. The algorithm for combining visibility and Level-of-detail to achieve interactive frame rates is described. Finally, conclusions are drawn and future directions listed.

VIEW-DEPENDENT ADAPTIVE TESSELLATION OF SPLINE SURFACES

We present a novel approach that combines the advantages of both backward and forward techniques in a unique way. We have been able to factor out sufficient computation to the pre-processing stage to make interactive adaptive view-dependent tessellation possible. This method generates detail only where necessary, thus reducing the total polygon count, has low memory overhead, allows arbitrary detail, and still is easy to implement. Our algorithm starts by pre-computing an object-space sampling for each surface element. It stores these domain points in a topologically sorted order so that the points that reduce more the deviation between the resulting triangles and the surface appear first. During rendering, the algorithm determines the samples that must be added or deleted to minimally achieve a user-specified screen-space deviation. It maintains a Delaunay triangulation [27] of the (two dimensional) domain samples, incrementally adding and deleting the desired samples. The triangles generated by mapping the domain samples to the surface form the approximation, and are sent to the graphics pipeline. If detail beyond the precomputed set of points is required, additional points are generated by uniformly sampling each domain triangle. We avoid tessellation cracks by choosing the same samples on both patches adjacent to the boundary curve. By speeding up the run-time sampling test and reducing the number of generated triangles, we are able to achieve a speed-up of about 2–4 over previous spline rendering methods [13, 14].

Algorithm

The algorithm consists of two main steps:

Pre-sampling: We compute a set of sample points on the domain. These sample points are 'good' in the sense that they are locally best at reducing the deviation of the resulting triangular approximation from the surface. These points also store the deviation of the approximation after the points are added. This information is used during rendering to decide which points to add.

View-dependent triangulation: At the rendering time, we start with a hierarchical organization of the pre-computed domain samples for each patch. We maintain a running Delaunay triangulation of the subset required for each frame. Based on the viewing parameters, we choose the samples that maybe deleted and others that must be added to insure a user-specified deviation in screen space is not exceeded. We incrementally update the

Delaunay triangulation. If more samples than have been precomputed are needed, we resort to uniform sampling [13] of the domain to add more samples. One challenge is to decide which samples to pre-compute and in what order to add or delete them to maintain coherence in triangulation.

Pre-Sampling

In principle, at the rendering time, we need to find:

1. which regions of the patch deviate more from the surface than desired, and which points, if added to the triangulation, would reduce the deviation to the desired value.
2. which points, if deleted, will leave the resulting triangulation still close to the surface.

The optimal answer is intractable. Even reducing the choice to a pre-selected set of points, while possible in our framework, requires long sampling and triangulation times: there are too many combinations of points to consider. We instead use a heuristic. For each patch, we compute the sorted list of samples as follows:

- Start with a minimal set of points. These points are always included in the approximation.
- Generate the Delaunay triangulation of this minimal set. This is our first approximation.
- While the deviation of the surface from the approximation is greater than a user specified tolerance 'Δ_0'
    1. Find the point on the surface that is farthest from the current approximation.
    2. Append the corresponding domain point to the list of pre-samples. Store the resulting deviation with the point.
    3. Insert the point in the Delaunay triangulation updating the current approximation.

At the end of the process, we have an ordered list, S, of domain samples for each patch and the object space deviation, D_i between the approximation and the surface if all points, S_j, $j \leq i$, are added to the triangulation for any *i*.

Deviation Computation

In order to find the point on the surface with the maximum deviation from the current approximation, we compute the maximum deviation of each triangle and use their maxima. In case a triangle is degenerate and lies on a straight line in the object space, we compute the unit vector parallel to it and the objective function simply computes the cross product of the displacement vector and this unit vector. In case all the three vertexes of the approximating triangle coincide, the function returns the norm of the displacement vector.

Note that a fixed order among the samples of a patch is acceptable for static tessellation, but the relative "importance" of points is view-dependent, and can vary for patches large on screen. We maintain a hierarchical partitioning of domains (a two level hierarchy is usually enough) into quads. As the projection of patch grows, we switch to a finer partition. We sample each sub-domain independently using a subsequence of patch pre-samples that lie on that sub-domain. The deviation for each domain is based on its own scale-factor (described in the next section).

Scale-Factor Computation

We first compute the maximum deviation of the perspective projection of the surface of a sphere from the projection of its center. Let **P** = (x, y, z) be the center of the sphere and let (ρ, θ, ϕ) be the local spherical coordinates of the vector starting at the center of the sphere which has the projection of maximum length. The global coordinates of the tip of the vector (say **Q**) are (x + ρsin(θ)cos(ϕ), y + ρsin(θ)sin(ϕ), z + ρcos(θ)). Let **XY** plane be the image plane and let (0, 0,—f) be the camera center.

Projection of **P** on the image plane = **P′** = (fx/(f+z), fy/(f+z), 0) and projection of **Q** = **Q′** = (f(x + ρsin(θ)cos(ϕ))/(f + z + ρcos(θ), f(y + ρsin(θ)sin(ϕ))/(f + z + ρcos(θ), 0). So maximizing the projection is equivalent to maximizing the length of **P′Q′,** which yields

$$|P'Q'| = \frac{fL^2}{(f+z)[(f+z)\sqrt{(L^2/\rho^2)-1} - \sqrt{x^2+y^2}\,)]}, \tag{2}$$

where $L = \sqrt{x^2 + y^2 + (f+z)^2}$.

This gives the ratio (γ) of length of the vector to its projection length = ρ / |**P′Q′**|. In particular amongst all the spheres of different radii ρ_i, the one that minimizes (γ) is obtained for a limit zero radius sphere, which yields

$$\lim_{\rho \to 0} \gamma = \frac{(f+z)^2}{fL} \tag{3}$$

We define scale-factor of a point **P** in object space as the length of the smallest vector anchored at P that projects to a unit vector in screen-space

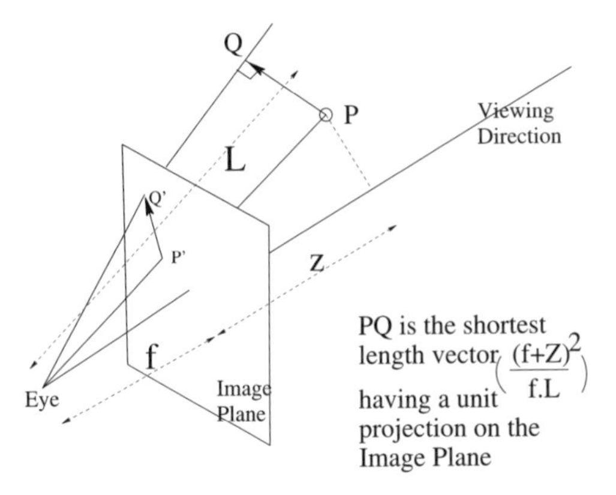

Figure 1. Scaling of a unit vector in screen space

(Figure 1). As computed above, this minimum value equals $(f+z)^2/(fL)$, where L is the length of the vector from the eye to the point **P**, and f is the focal length, and (f + z) is the length of the projection of the vector from the eye to the point **P** , along the principal viewing direction.

Delaunay Triangulation

We use the algorithm by [28] to perform incremental Delaunay triangulation. The algorithm is outlined below.

To insert a given point *p* in a given triangulation Δ, we need to find t_i, the set of triangles whose circumcircle contain *p*. We perform this search by walking. We retain the centroid C, of Δ, which serves as the starting point of this walk. Searching proceeds as follows:

- Find a consecutive pair of edges, (e_{j1}, e_{j2}) of triangle t_j, such that *p* lies in the positive half space of (e_{j1} and e_{j2}).
- If *p* is contained in the circumcircle of t_j, find one of the intercepted triangles.
- Otherwise, continue at another edge pair adjacent to one of the vertexes of t_i.

If no intercepted triangle is found, we simply connect *p* to the visible vertexes of the convex hull of current triangulation. If an intercepted triangle t is found, we perform a breadth first traversal of the adjacency graph to discover t_i, all intercepted triangles, whose overall boundary forms a star shaped polygon **P**. We delete all t_i, and add an edge between *p* and all the vertexes of **P**. The centroid vertex is also updated at this stage. Storing the centroid vertex reduces the expected search time to O ($n^{0.5}$) though the worst case remains linear. The insertion procedure takes O (k), where k is the number of edges in **P**.

To delete a point *p*, we first locate it in the triangulation. We delete all triangles adjacent to *p* thus obtaining again a star shaped polygon **P**. If **P** has more than three edges, we insert each edge of **P** in a priority queue. The priority value for an edge $e_i = r^2_i - (|c_i - p|^2)$ where c_i is the circumcenter of the triangle formed by edge e_i and e_{i+1}, and r_i is its radius. We then add the triangles to Δ in the following order:

1. Choose the edge, e_i with the minimum priority and add the triangle formed by e_i and e_{i+1} to Δ.
2. Replace e_i and e_{i+1} from **P** and add the new edge.
3. Repeat until **P** is a triangle.

In the end, we update the centroid vertex. The expected search and actual deletion times are similar to the one's obtained during insertion.

Render-time Sample Selection

Our rendering algorithm allows the users to bound the maximum geometric deviation of the triangular approximation. This bound may be specified in pixel units. If the user-specified screen space bound for a patch B is d_s, object-space deviation required for the approximation is $d_o = s(B)d_s$, where s(B) is the scale-factor for patch B.

Different patches, indeed different points of a patch, may have different scale-factors. We use an octree based spatial partitioning of space. For all patches contained in a sufficiently small partition, we use the same scale factor. Typically, partitions closer to the viewpoint are more refined than those further away, as the scale factor closer to the viewpoint varies faster. The sample selection proceeds as follows:

- Start with the octree cubes used in the previous frame. We call a cube *terminal* if the scale-factor of eight corners of the cube differ by less than δS, a user-specified tolerance. If a cube is non-terminal, we subdivide it. Otherwise, if a cube's parent is terminal, we recursively use the parent. If δS is chosen to be $(1/d_s)$, the approximation does not under-deviate by more than a pixel.
- The scale-factor of a cube is the smallest of the scale-factors of its corners. For each patch completely contained in a terminal cube, C, with scale factor s(C), we choose all samples S_i, $i \leq j$, such that the associated deviation $Dj < s(C)$ and $D_{j-1} > s(C)$.
- Suppose the value of j in previous frame is j_{prev}, and in the current frame

is j_{curr}. If $j_{curr} > j_{prev}$, we add samples $S_{jprev+1}..S_{jcurr}$ to the Delaunay triangulation, otherwise we delete $S_{jcurr+1}..S_{jprev}$.

- If a patch lies in more than one terminal cube that are all adjacent to each other, the minimum scale-factor of those cubes is assigned to that patch.

Uniform Dynamic Sampling

If the required deviation $d_0 < D_n$ for a patch with n precomputed samples, we compute more samples in a lazy fashion. We do this render-time sampling using a variant of the uniform sampling method of Kumar, et al. [13]. We store, for each triangle t of patch B, $t_u = 1 / \sqrt{|B_{uut}|_{max}}$ and $t_v = 1 / \sqrt{|B_{vvt}|_{max}}$. At rendering time, the uniform step sizes in u and v directions are given respectively by $k_d t_u \sqrt{s(B)d_s}$ and $k_d t_v \sqrt{s(B)d_s}$ with constant k_d normally set to 1 (see [14] for details).

Crack Prevention

If we independently tessellate two adjacent patches, we could choose different samples on their common boundary. This results in a crack in the resulting approximation. In our approach, we want to keep the tessellation of adjacent patches independent of each other to facilitate easy parallelization. We assume that the object space representation of the boundary curves are the same. We pre-compute the sampling of each boundary curve separately from the interior. Ensuring that the same samples are used at the boundary of both adjacent patches to it eliminates cracks.

Implementation and Results

We have implemented our algorithm and tested it on a variety of models. All timings reported in this paper are from an Onyx2 with a 195MHz R10K Infinite Reality graphics. Our experiments consisted walkthrough of a variety of models. We first compare the number of triangles produced by our algorithm with that by a uniform tessellation algorithm [13] in Table 1. This combined with faster sample selection, results in the overall speedup, which is more pronounced for large models. In Table 2, we show the render-time behavior of our algorithm. The number of samples that need to be added or deleted from the triangulation is less than 1% of the average triangle count, and the overhead of the Delaunay triangulation is less than 10% of overall time. Thus, per frame operation is quite efficient. In addition, uniform tessellation is needed for less than 0.15% of the patches on an average. Table 3 reports the pre-processing time. In addition, it lists the time it would take to pre-compute the three dimensional position and normal values. Note that the number of pre-samples is rather large, but we

Table 1. Comparison of the number of triangles produced by our algorithm v/s [13] for the same screen space deviation of two pixels

Model	Num Patches	Avg num of Tris/frame	
		Ours	Kumar et al.
Goblet	72	2302	6744
Coke	330	3994	5488
Dragon	5079	16804	42115
Garden	38646	83733	122360

Table 2. Run-time behavior of our algorithm

Model	Number of sample updates/frame	Time to update Triangulation	Overall rendering frame-rate	Rendering rate of [13]
Goblet	24	.11 ms	72	42
Coke	53	.28 ms	65	23
Dragon	653	9 ms	18	5
Garden	1485	16.2ms	6	1.7

Table 3. Pre-Sampling performance

Model	Num Samples pre-computed	Pre-process time	If Vertex/Normal are pre-computed
Goblet	6,500	48.78 Seconds	0.41 Seconds
Coke	23,109	58 Seconds	.45 Seconds
Dragon	864,610	31 minutes	38 Seconds
Garden	838,152	13 minutes	14.1 Seconds

keep only eight bytes per sample, including the multiplication factor needed for uniform sampling.

POINT BASED RENDERING FOR PARAMETRIC SURFACES

On carrying out extensive experiments, we discovered that surfaces that are small on the screen space are better approximated as squares (or circles), than triangles. Point based approximation generates better quality at a fraction of rendering cost. Our experiments show that ensuring less than a single-pixel deviation for a patch 100 pixels on the screen requires 2.5–3 times more points than triangles. However, the triangle based rendering is half as fast due to a much larger overhead. Our algorithm for point and triangle based tessellation is based on a simple idea. We pre-compute and store an ordered list of samples for each surface. At rendering time, we simply choose the correct samples from the list. The challenge is to determine which samples to store and which samples to use for the given viewing parameters. In the following section, we describe the algorithm for point-based tessellation.

Spheres as elements

Imagine spheres at the sampled points on the surface such that every point on the surface lies inside at least one sphere. In other words, when these spheres are projected on the image plane, the skewed ellipses thus formed would have no holes between them (Figure 2). In contrast, the space around the samples is traditionally filled by interpolating triangles over the surface. As we prove later, the spheres centered on these sampled points have the property that the actual surface does not deviate by more than the radius of the sphere from the surface of the sphere. We choose the sample points so that they locally reduce the deviation of the surface from the approximating surface (i.e. the surface of the sphere), and hence reduce the radius of the sphere. To render an individual sphere, we compute the center C, of the projected ellipse (by projecting the sample point). We next compute the maximum deviation (say d) of the elliptical surface from C. Now if we render a square splat centered at this point of dimension $2d$ centered at C, the whole surface of the projected ellipsoid is covered (Figure 3). The color and normal value assigned to each of the $4d^2$ pixels are the color and normal values of the original point. At run-time, we need to compute C and d. As explained later, assigning the same value of d to all spheres on a spline patch greatly simplifies the problem without increasing by much the total number of spheres needed for a given error bound. We will prove that the value of d we assign to each patch is optimal up to integral point sizes. In other words, the value of d we compute is no more than 0.5 pixel away from the optimal value. Thus for each patch, we choose samples that when projected as points on the screen, does not leave any holes, and also obeys the deviation bounds. To compute the samples and their point sizes, we just need log M lookups for a patch, where M is the total number of precomputed samples.

Spheres on the patch in Object Space

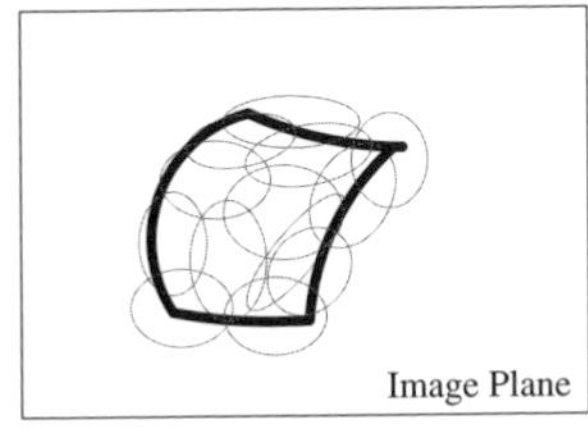

Projection of Spheres with no holes

Figure 2. Projection of surface samples on Image Plane

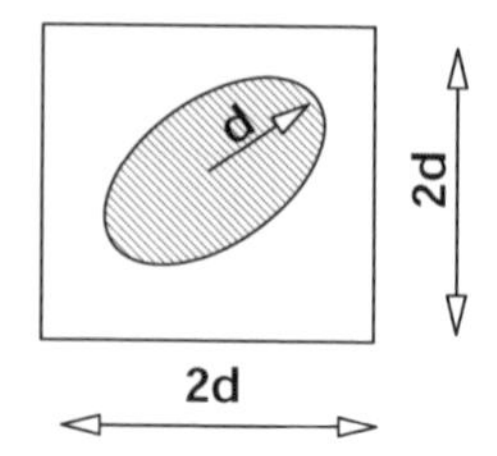

Figure 3. Square splat covering the projection of a sphere on the image plane

Algorithm

The two main steps of the algorithm are:

1. Pre-Sampling

In principle, at rendering time, we need to find:

1. which regions of the patch deviate more from the surface than the desired value, and which points if added, would bring down the deviation to the required level without introducing any screen-space holes, something that we take care of in object space itself, unlike some other suggested algorithms [24], which perform an image space interpolation to fill up the holes that these points might introduce.
2. which points, if deleted, will leave the resulting points satisfying the properties of bounded deviation and no screen-space holes.

We compute the sorted list of samples as follows:

1. Start with a minimal sample set (e.g. the four corners of the domain). These points are always included in the approximation.
2. Generate the delaunay triangulation of the minimal set. Now compute the center and radius of the circumscribing spheres for each of the triangles obtained.
3. While the sphere with the maximum radius has a radius greater than a user specified tolerance Δ_0,
    - Choose the triangle (in domain space) that has the maximum circumscribing radius r_{max}, and let R be its center.
    - Append R to the list of pre-computed samples and also the corresponding radius r_{max}.
    - Insert the point in the Delaunay triangulation updating the current approximation.

At the end of the process, we have an ordered list, S, of domain samples for each patch, and also the maximum radius amongst the spheres which covered the surface before adding that particular point. We now claim the following two statements, and provide simple proofs, which show how this process of adding points gives us samples, which cover the whole surface without leaving any gaps, and have bounded deviation at the same time.

Claim 1: *Maximum deviation of a surface patch from the approximating sphere is equal to the radius of the sphere that encloses that patch.*
Proof: The proof follows simply from our methodology of selecting the radius of the sphere which encloses the whole sub patch corresponding to the triangle on the domain: every point inside a sphere is at a distance less than or equal to radius of the sphere from its surface.

Claim 2: *At any instant of the domain triangulation, let r_{max} be the maximum radius amongst the circumscribing spheres. If we draw spheres with radius = r_{max} on all the sampled points on the surface, there would not be any holes on the surface.*

Proof: At any instant of the domain triangulation, each triangle has a circumscribing radius less than or equal to r_{max}. Now, for any triangle, if r is the radius of the circumcircle, then if we draw circles with centers as the three vertexes, and the same radius, no point on the triangle would be left uncovered (Figure 4). Hence drawing spheres with a radius greater than r (r_{max}, to be specific), would not leave any gaps on the surface of any of the triangles that share a particular vertex, and hence none on the surface comprising these points.

Moreover, the maximum deviation of the surface from these spheres still remains the same, because for each circumscribing sphere of a triangle, each surface point is at most at a distance of *radius* from the center and each of the three substituting spheres pass through the center of the original

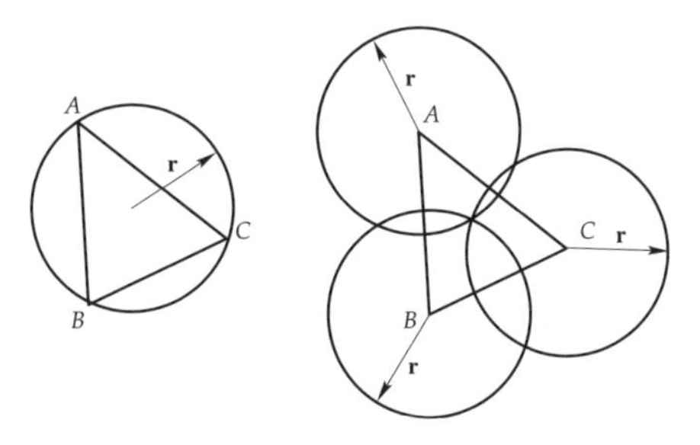

Figure 4. Covering surface by drawing spheres at the vertices

sphere. Hence selecting points in the above fashion simultaneously satisfies the deviation criteria and the criterion for no holes. Also with each sample point, we get the maximum object space deviation D_i (for the ith selected point), between the approximation and the surface if all the points S_j, $j \leq i$ are rendered. Note that $D_{i+1} \leq D_i$[1] and $D_{|S|} \leq \Delta_0$. Thus given a deviation bound d_0, we can find the prefix of S that generates an approximation with deviation less than or equal to d_0.

Computation of Sphere Parameters

For each triangle in the object space, we need to find the center and radius of the sphere, which encloses every point on the surface corresponding to the region inside the triangle on the domain space. Let the three domain points be represented as $t = (p_1, p_2, p_3)$, $p_i = (u_i, v_i)$, of the Bezier surface patch **B**. Let $P_1 = \mathbf{B}(p_1)$; $P_2 = \mathbf{B}(p_2)$ and $P_3 = \mathbf{B}(p_3)$ be the corresponding object space points. Now compute the circumcenter, say Q, and the radius of the circumcircle, say r_1. The following three possibilities are given special values.

1. All the three points are coincident. So $Q = P_1$ and $r_1 = 0$.
2. The three points lie on a straight line. Let P_1 and P_2 be the extreme points. So $Q = (P_1+P_2)/2$ and $r_1 = |P_1P_2|/2$.
3. The three points form an obtuse angled triangle. Let P_1 and P_2 be the end points of the longest edge. So $Q = (P_1+P_2)/2$ and $r_1 = |P_1P_2|/2$.

Shoot a ray QR, perpendicular to the plane of the points, intersecting the surface at point R. We use Powell's method [29] to find the point of intersection. Let $|QR| = r_2$. Imagine a sphere with center R and radius $r = \sqrt{r_1^2 + r_2^2}$ and check whether all surface points corresponding to the triangular domain (p_i for $i \in (1, 2, 3)$) lie within the sphere (Figure 5). In case any of them lies outside, increase the radius so that finally the whole sub patch lies within the sphere.

2. Render-time sample selection

At run-time, to guarantee a given frame rate, we can render only a certain number of points per frame on a given graphics platform. Let us say the maximum number of allowable points per frame is C, a user specified

1. Technically, deviation could increase on adding a new point for some degenerate patches. Even in such cases, adding a sequence of samples always leads to a lowering of deviation. We add or delete such samples together and assign a single index to them.

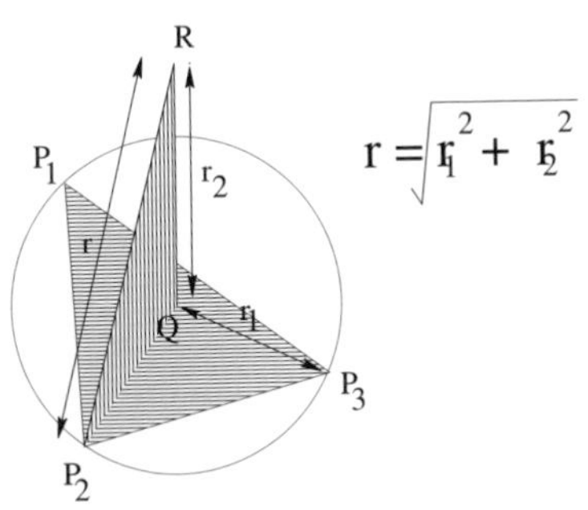

Figure 5. Computation of radius and center of the circumscribing sphere

constant, or which can be determined for any system by running some experiments on test cases which give an estimate of the maximum number of points that can be rendered per second. We need to distribute these C points among different patches of the scene. Let the total number of non-planar patches be N (the planar patches can be rendered as *two* triangles each, with no need for point rendering). For distribution purpose, we would want to minimize the screen space error for every patch. Hence our error metric is the maximum screen space error among all the patches, and we want to minimize this value. As shown in the next section, we compute a solution that gives the optimal screen space error (up to integral point sizes), and gives an error of at most a unit point size (w.r.t. non integral point sizes). In practice, rendering integral point sizes is comparably much faster than the other alternative. Hence our algorithm gives the optimal point sizes to be used for rendering each point on any particular patch. For the rest of the subsection, assume we have allocated a point size of d^j_i for the jth frame to the ith patch, for $i \in [1...N]$. In general, a point size of d is equivalent to stating that the maximum screen space deviation of the actual surface from the approximation surface is equal to $d_s = d/2$. This bound is in pixel units and the object-space deviation required for the approximation is $d_0 = s(B)d_s$, where s(B) is the scale-factor for the patch B (explained in the previous section).

Point Distribution

The point distribution problem needs special attention, as we need to update it every frame, and hence should spend as little time as possible on it. The mathematical formulation of the problem is as follows:

Let us say the frame in question is the jth frame, and the scale-factor for the ith patch is s^j_i . Hence, if a point size of δ^j_i is chosen, we can compute the maximum allowable object space deviation for the patch ($= s^j_i\, \delta^j_i /2$). This object space deviation equals the maximum allowable radii of spheres on the surface. So we need to lookup the list of pre-computed values to compute the exact number of spheres required. This requires at most $\log(M_i)$ lookups, where M_i is the number of pre-computed samples for the ith patch. So we can define a function f^j_i: R→N, which takes the point-size of the rendered points for that patch, and returns the number of points required. We can obtain such f^j_i for all $i \in [1...N]$.

Hence the optimization problem can be stated as follows:

$$\text{Minimize } (\text{Max}_i\ \delta^j_i)$$
$$\text{s.t. } \Sigma_i\ f^j_i\ (\delta^j_i) \leq C$$
$$\text{and } \delta^j_i > 0.$$

f^j_i resembles a step function (see Figure 6), and we need to formulate this function for every patch for each frame. It is clear that the optimal solution would assign the same point size to each patch. Solving the above equation analytically, to get the optimum solution, might require in the worst-case $O(N(M_{max}))$ steps, where M_{max} is the maximum number of pre-computed points for any patch. This is clearly an expensive solution, and also overkill, as in practice, we need integral size solutions. So we propose the following algorithm, which does not require computing the function f^j_i for the whole point size range, but instead just for a few discrete values.

1. Assign a point size of p′= 1 for each patch.
2. Compute the number of total points required, and let $C' = Ó_i C_i$ where C_i is the number of points required for the ith patch.
3. If $C' \leq C$, then render all the patches with this point size.
4. If $C' > C$, increase the point size p′ by 1, and go back to step 2.

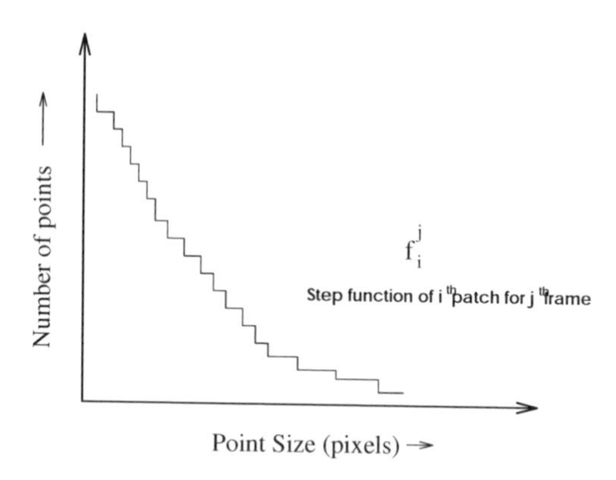

Figure 6. Step Function for a particular position of some patch

The above algorithm does a linear search (can be easily replaced by binary) to arrive at a point size that satisfies the budget. Clearly any integral point size less than p′ would overshoot the budget. However, this algorithm requires log K steps to arrive at the correct point size, where K is the maximum point size allowed. In order to improve it, we exploit the temporal coherence of the movement of the navigator to get a 2n-time bounded approximation algorithm.

1. Allocate each patch the point size it had in the previous frame.
2. Compute the number of total points required, and let $C' = Ó_i C_i$ where C_i is the number of points required for the ith patch.
3. If C′ < C, decrease the point sizes of each of the patches by one, and recompute C′ after processing each patch, and terminate when C′ > C.
4. If C′ > C, increase the point sizes of each of the patches by one, and recompute C′ after processing each patch, and terminate when C′ < C.

In practice, because of the temporal coherence of the eye point, the solution does not change much between frames, and hence usually in at most 2 passes, we obtain the optimal integer solutions.

Claim 3: *If p′ is the optimal solution, we obtain p′ as our solution, and hence the screen space error is within 0.5 pixel of the optimal solution.*

Claim 4: *Let p_i and p_j be the point sizes allocated to patches i and j respectively. Then* $|p_i - p_j| \leq 1$.

Proof: The proof follows simply from our methodology of changing point sizes for each patch. Each patch starts with a point size of 1 (before the first frame), and they are sequentially increased or decreased in round-robin order. Hence the difference in point sizes between any two patches would at most be one.

The above two claims show that our simple algorithm gives us good results while taking minimal processor time. For a better bound on the point size, we could carry out a binary search, using the same methodology. However, the cost of computing a tighter approximation far exceeds the benefit of using such a solution.

Implementation and Results

We have implemented our algorithm and tested it on a variety of models. All timings reported in this paper are from an Onyx2 with a 400 MHz R10000 and an InfiniteReality graphics card. Our experiments consisted of

Table 4. Pre-Sampling performance

Model	Number of Patches	Num Samples Pre-computed	Pre-process time in minutes
Teapot	32	129273	10
Spoon	66	234,290	19
Goblet	72	123,396	17
Dart	100	141,150	09
Coke	330	610,354	44
Scissors	505	141,243	14
Pencil	570	2,116,616	172
Dragon	5354	20,269,877	1664
Garden	38646	8,842,370	804

Table 5. Run-time behavior of the algorithm

Model	Points per frame	Average error	%-time spent in software	Frame rate
Teapot	90,000	1.8	0.3	31
Spoon	90,000	1.1	0.6	34
Goblet	100,000	1.45	0.5	34
Dart	90,000	0.7	0.6	36
Coke	90,000	2.09	0.9	25
Scissors	90,000	1.7	0.9	24
Pencil	70,000	3.00	2.44	23
Dragon	50,000	3.12	11.9	20
Garden	50,000	7.5	19.1	7

viewing a variety of models from various viewpoints. In Table 4, we report the pre-processing time for different models. This pre-processing algorithm takes time that is proportional to the total surface area of the model, and not the number of patches. Hence some of the models with large number of patches can be pre-processed comparatively faster. The pre-processing for any patch is independent from any other patch of the model. Hence pre-computation can also be carried out in parallel. However, all the times reported are for a single processor machine. Also, we compute a large number of samples per patch (by giving a small deviation threshold), to carry out extensive tests. In practice, for a rendering capacity of C points per frame, around 10C points should suffice.

In Table 5, we report the time spent by the algorithm to compute the appropriate point samples that need to be sent to the graphics pipeline. Also, we report the average screen space error for the simulated browsing of the models. It can be seen that a very small fraction of total rendering time (about 10%) is spent in software to figure out the correct samples. We achieve real time frame rates for most of the models on a hardware customized for triangle rendering. As we spend very little of the total rendering time in determining the basic primitives to be rendered, we believe that our algorithm would give much better results on hardware customized for point rendering. It is worth mentioning that we have just incorporated basic view frustum culling. We can also use backface culling [30], to potentially compute a tighter bound to the number of contributing patches per frame. Another interesting observation was that when the screen space area of a patch was considerably large, screen space errors of even 3–4 pixels did not produce any noticeable artifacts. This can be explained by the small percentage of error in the projection. Hence, we can use the metric of relative screen-space error (obtained by dividing the screen-space error with some normalized area of the screen space projection). We are still working out the details of this metric, and haven't mentioned it in the results.

COMBINING VISIBILITY AND LEVEL OF DETAIL (vLOD)

In this section, we describe our framework that combines both simplification and visibility computation at the same time. We present our algorithm for models specified in form of polygons. However, this framework is applicable to any class of input models, including parametric surfaces, something we have focused on in this research.

vLOD Management

In this section, we describe the computation and maintenance of vLOD's in our framework. The basic concept of vLOD is quite simple:

1. Hierarchically partition the viewing space.
2. For each partition, or cell, adapt the level of detail so that the screen-space error when viewed from anywhere in the cell is bounded (using any geometric error-based simplification algorithm). Given a screen-space error bound, we present the procedure to compute the worst-case object space error of all the viewpoints in the convex cell.
3. Obtain vLOD by discarding the geometry that is guaranteed not to be visible from anywhere in the cell (using any cell-based visibility algorithm).
4. Encode and compress the differences Δ's) between vLOD of a cell and its neighboring cells.
5. At rendering time, construct vLOD of the cell the viewer is in (using Δ's) and display it.

The differences are small on average, making the vLOD framework possible. To make it really practical, though, we must manage and encode the Δ's in such a way that storage requirements are reduced, and the time to reconstruct vLOD's from them is very small. The strength of vLOD framework lies in its simplicity and its ability to use any LOD adaptation and visibility algorithm. Our prototype implementation works well on commodity graphics hardware. It scales well and is able to interactively render models larger than available memory. vLOD also supports bounded screen space error as well as guaranteed speed visualization (though, not both at the same time).

Spatial Partitioning

Ideally, our view-space partitioning has three requirements:

- Visibility and detail of objects as seen from different points inside a single cell do not vary much. This ensures that we do not use more geometry than is ideally needed for the given viewing parameters.
- The vLOD's of adjacent cells do not vary by much. This ensures that the size of D's is small. Visibility sometimes changes drastically between cells. Hence the worst-case size of D can be large although the average size is small.
- The number of cells are small and their vLOD's may be easily computed. This ensures small storage overhead and efficient pre-processing.

In our implementation, we use an octree-based partitioning scheme. We start with a uniform grid and refine it further. A cell, C, is too large and hence subdivided, if its vLOD(C) is larger than a user specified value. Occasionally, subdividing these cells does not produce sufficient reduction in vLOD's. In this case, we check the ratio of its vLOD and $vLOD_{min}(C)$, where $vLOD_{min}(C)$ is the smallest $vLOD(C_i)$, for all subdivided cells, C_i of C. If vLOD(C) is no more than twice $vLOD_{min}(C)$, we do not subdivide the cell C.

vLOD Construction and Cell Differences

In this section, we describe our technique for computing and storing vLOD's assuming a visibility and simplification algorithm. Because of lack of space, the specific details of these algorithms will not be discussed, but for now, it suffices to assume that the visibility algorithm produces a list of geometry identifiers that are visible from a given cell (we call these visible

IDs). Similarly, we assume that the simplification algorithm produces a list of LOD IDs, a list of identifiers for geometry in the LOD of a given cell. Each Visible ID that is also an LOD ID is added to the vLOD. We find the differences in the list of IDs and encode these.

Implementation and Results

We have implemented our vLOD system and tested it on a city model consisting of 52.4 million triangles. We have implemented on SGI Onyx2 and report our results on this platform. The images generated are the same if the visibility and the simplification were computed at rendering time. Our method doesn't discard any invisible geometry and uses an appropriate level of detail for visible geometry. We have performed experiments to check the coherence. In typical walkthroughs, one only needs to cross a boundary in less than 2.5 % of the frames. We use asynchronous reads to prefetch the relevant Δ's from the disks. The average size of Δ is around 20 % of the vLOD size of a cell. Additionally, Δ(C) is more than half the size of vLOD(C) in less than 10 % of the frames.

CONCLUSIONS AND FUTURE WORK

We have presented the first view-dependent adaptive spline tessellation algorithm and also an algorithm to render parametric patches using points as the basic primitives. These algorithms do most of the computation offline. At rendering-time, they just perform minimum computation to select the set of samples that need to be rendered. We give guaranteed frame rates or error bounds to the user. For patches too close to the viewpoint, we use triangles. Pixels that are smaller than a pixel in size are grouped and rendered as single points. We are able to obtain real-time rendering rates with small errors for most of the models. We continue to refine our implementation. Current hardware is often not well optimized for rendering points. We expect the gains of using point based tessellation to increase when such optimization becomes routine. We are also extending our algorithm to trimmed splines and subdivision surfaces. In addition, we have also presented a novel way to combine rendering acceleration techniques into a common framework. In fact, as more efficient simplification and visibility algorithms are developed, they can be easily incorporated into our system. All these schemes would aid in real-time interaction with extremely large spline surfaces.

A lot of future directions of research are possible. For the view-dependent adaptive tessellation algorithm, we have used the incremental Delaunay scheme. More investigation is needed into other competing triangulation schemes. Also, during the point rendering algorithm, there is a noticeable difference between triangle based and point based tessellation when a switch is made between the two. We are working on schemes to smooth the transition between two.

ACKNOWLEDGMENTS

I would like to thank my advisor Dr. Subodh Kumar, Dr. Shankar Krishnan, Dr. Jonathan Cohen and Budirijanto Purnomo for insightful discussions. The models were courtesy of Lifeng Wang, the modeling group at University of British Columbia and XingXing Graphics Co. (Garden), David Forsey (Dragon) and Alpha 1 System (Soda can and Goblet). This work was supported by the Link Foundation.

REFERENCES

[1] "id Software, Inc." *Web Page: http://www.idsoftware.com*
[2] "Remedy Entertainment" *Web Page: http://www.remedy.fi*
[3] G. Farin, "Curves and Surfaces for Computer Aided Geometric Design: A Practical Guide" *Academic Press Inc.,* (1993).
[4] J. Whitted, "An improved illumination model for shaded display" *Siggraph 1979 Proceedings,* 13(3): 1–14, (1979).
[5] J. Kajiya, "Ray tracing parametric patches" *SIGGRAPH 1982 Proceedings,* 6(3): 245–254, (1982).
[6] T. Nishita, T. Sederberg and M. Kakimoto, "Ray tracing trimmed rational surface patches" *Siggraph 1990 Proceedings,* 24(4): 337–345, (1990).
[7] E. Catmull, "A Subdivision Algorithm for Computer Display of Curved Surfaces" *Ph.D. thesis, University of Utah,* (1974).
[8] M. Shantz and S. Chang, "Rendering trimmed NURBS with adaptive forward differencing" *Siggraph 1988 Proceedings,* 22(4):189–198, (1988).
[9] J. Whitted, "A scan line algorithm for computer display of curved surfaces" *Siggraph 1978 Proceedings,* 12(3): 8–13, (1978).
[10] J. Blinn, "Computer Display of Curved Surfaces" *Ph.D. thesis, University of Utah,* (1978).
[11] J. Lane, L. Carpenter, J. Whitted and J. Blinn, "Scan line methods for displaying parametrically defined surfaces" *Communications of ACM,* 23(1): 23–34, (1980).
[12] S. S. Abi-Ezzi and L. Shirman, "Tessellation of curved surfaces under highly varying transformations" *Eurographics Proceedings,* 385–397, (1991).
[13] S. Kumar, D. Manocha and A. Lastra, " Interactive display of large NURBS models" *IEEE Transactions on Visualization and Computer Graphics,* 2(4): 323–336, (1996).
[14] S. Kumar, D. Manocha, H. Zhang and K. Hoff, "Accelerated walkthrough of

large spline models" *Symposium on Interactive 3D Graphics Proceedings,* 91–101, (1997).

[15] D. Filip, "Adaptive Subdivisin algorithms for a set of Bezier triangles" *Computer-Aided Design,* 18(2): 74–78, (1986).

[16] V. Vlassopoulos, "Adaptive Polygonization of parametric surfaces" *Visual Computer,* 6: 291–298, (1990).

[17] D. Aliaga and J. Cohen et. al., " MMR: An integrated massive model rendering system using geometric and image-based acceleration" *Proc. Symposium on Interactive 3D Graphics,* 101–106, Atlanta, GA, (1999).

[18] J. Cohen, A. Varshney, D. Manocha and G. Turk et. al., "Simplification Envelopes" *Siggraph 1996 Proceedings* ,119–128, (1996).

[19] M. Garland and P. Heckbert, "Surface Simplification using quadric error metrics" *In Siggraph 1997 Proceedings,* 209–216, (1997).

[20] M. Levoy, et. al., "The Digital Michelangelo Project: 3D Scanning of Large Statues" *Siggraph 2000 Proceedings,* 131–144, (2000).

[21] M. Levoy and T. Whitted, "The Use of Points as a Display Primitive" *Technical Report TR-85–022,* University of North Carolina, Chapel Hill, (1985).

[22] J. Grossman and W. Dally, "Point Sample Rendering" *Eurographics Rendering Workshop Proceedings,* (1998).

[23] S. Rusinkiewicz and M. Levoy, "QSplat: A Multiresolution Point Rendering System for Large Meshes" *Siggraph 2000 Proceedings,* 343–352, (2000).

[24] H. Pfister, M. Zwicker, J. Baar and M. Gross, "Surfels: Surface Elements as Rendering Primitives" *Siggraph 2000 Proceedings,* 335–342, (2000).

[25] M. Wand, M. Fischer, I. Peter, F. Heide and W.Straber, "The Randomized z-Buffer Algorithm: Interactive rendering of Highly Complex Scenes" *Siggraph 2000 Proceedings,* 361–370, (2000).

[26] J. Cohen, D. Aliaga and W. Zhang, "Hybrid Simplification: combining Multi-resolution Polygon and Point Rendering" *IEEE Visualization 2001 Proceedings,* (2001).

[27] L. DeFloriani and E. Puppo, "An online algorithm for constrained Delaunay triangulation" *Computer. Vision Graph. Image Processing,* 54:290–300, (1992).

[28] O. Devillers, "Improved incremental randomized delaunay triangulation" *14th Annual ACM Symposium on Computational Geometry Proceedings,* 106–115, (1998).

[29] W. Press, S. Teukolsky, W. Vetterling and B. Flannery, "Numerical Recipes in C: The Art of Scientific Computing" *Cambridge University Press,* (1993).

[30] S. Kumar, D. Manocha, W. Garrett and M. Lin, "Hierarchical Back-Face Culling" *Computers and Graphics,* 23(5): 681–692, (1999).

Evaluation and Application of Algorithms For a Hybrid Environment System

Benjamin C. Lok

Department of Computer Science

University of North Carolina at Chapel Hill

Chapel Hill, NC 27599–3175

Research Advisor: Dr. Frederick P. Brooks, Jr.

ABSTRACT

Suppose one has a virtual model of a car engine and wants to use an immersive virtual environment (VE) to determine whether both a large man and a petite woman can readily replace the oil filter. This real world problem is difficult to solve efficiently with current modeling, tracking, and rendering techniques. Hybrid environments, systems that incorporate real and virtual objects within the VE, can greatly assist in studying this question. In this paper we describe new algorithms for generating virtual representations, *avatars*, of dynamic real objects at interactive rates and enabling virtual objects to interact with and respond to the real-object avatars. This allows dynamic real objects, such as the user, tools, and parts, to be visually and physically incorporated into the VE. The algorithms use image-based object reconstruction and a volume-querying mechanism to detect collisions and to determine plausible collision responses between virtual objects and the real-time avatars.

We then evaluate the algorithms from various standpoints:

- (Engineering)—We present an implementation of the reconstruction and collision detection algorithms in a prototype system
- (Theoretical)—We conduct performance and error analysis for the algorithms.
- (Usability)—Beyond theory though, are hybrid environments even practically useful for VE tasks? We conducted a user study that evaluated the hybrid environments' effect on VE task performance and sense-of-presence.
- (Applicability)—We looked to evaluate hybrid environments in the context of a real-world task.

The study showed that for spatial cognitive manual tasks, hybrid environments provide a significant improvement in task performance measures. Also, participant responses show promise of improving *sense-of-presence* over customary VE rendering and interaction approaches.

We detail our beginning collaboration with NASA Langley Research Center to apply the hybrid environment system to a satellite payload assembly verification task. In an informal case study, NASA LaRC payload designers and engineers conducted common assembly tasks on payload models. The results suggest that hybrid environments could provide significant advantages for assembly verification and layout evaluation tasks.

INTRODUCTION

Motivation

Conducting design evaluation and assembly feasibility evaluation tasks in immersive virtual environments (VEs) enables designers to evaluate multiple designs more efficiently than if mock-ups are built and more thoroughly than can be done from drawings. Design review has become one of the major productive applications of VEs [1]. Virtual models can be used to study: 1) can an artifact readily be assembled? and 2) can repairers readily service it? The ideal VE would be visually identical to the real task. In the assembly verification example, parts and tools would have mass, feel real, and handle appropriately. The participant would naturally interact with the virtual world, and in turn, the virtual objects would respond to the participant's action appropriately [2].

Obviously, current VEs are far from that ideal system. Indeed, not interacting with every object as if it were real has distinct advantages, as in dangerous or expensive tasks. In current VEs, almost all objects in the environment are virtual, but both assembly and servicing are hands-on tasks. The principal drawback of virtual models—that there is nothing there to feel, to give manual affordances, and to constrain motions—is a serious one for these applications. Simulating a wrench with a six degree-of-freedom wand, for example, is far from realistic, perhaps too far to be useful. Imagine trying to simulate a task as basic as unscrewing an oil filter from an engine in such a VE!

Getting shape, motion, and inputs from real objects requires specific development for modeling, tracking, and interaction. Every possible input, action, and model for all objects, virtual and real, needs to be defined, developed, and implemented.

Also the visual representations of these objects within the VE, their *avatars,* are usually stylized and not visually faithful to the object itself. We extend our definition of avatar to include a virtual representation of *any* real object. Ideally, these real-object avatars are registered in look, form, and function with the real object.

We believe a *hybrid environment* system, one that could handle *dynamic real objects,* would be effective in providing natural interactivity and visually-faithful self-avatars. We define *dynamic objects* as real objects that can change shape and appearance. We define *incorporating real objects* as generating avatars—registered with their real object counterpart—that interact with purely virtual objects. Such a system would allow designers to see if there is enough space to reach a certain location or train people in assembly with real parts, tools, and handling the physical variability among participants.

Approach

First, we developed a hybrid environment system that uses camera-based reconstruction algorithms to generate real-time virtual representations of real objects. Next, we developed algorithms to use the virtual representations in virtual lighting and in physically-based mechanics simulations. In a sense, there is a merging of two spaces, the physical space (real objects) and virtual space (corresponding virtual objects). The participant sees, handles, and feels real objects while interacting with virtual objects.

Then, we looked at applicability by conducting a user study on whether hybrid environments provide any benefit for typical VE tasks. The user study results show a statistically significant improvement in task performance measures for interacting with real objects within a VE compared to interacting with virtual objects.

Finally, we looked at the usability of the system in a case study on applying the hybrid environment be applied to a real world task. The participants' experiences anecdotally showed the effectiveness of handling real objects while interacting with virtual objects.

PREVIOUS WORK

Incorporating Real Objects into VEs

Modeling. Prebuilt models are usually not available for specific real objects. Making measurements and then using a modeling package is laborious for complex static objects, and near impossible for capturing all the degrees of freedom of complex dynamic objects.

Automated capture systems assist in generating models from real objects. The 3–D Tele-Immersion work uses dense-stereo algorithms to create virtual representations of participants for tele-communication applications [3]. Matusik, *et al.*, presented an image-based visual hull algorithm, "Image Based Visual Hulls" (IBVH), that uses image-based rendering to calculate the visual hull at interactive rates. Their work also provides methods to compute visibility, coloring, and polygonal meshes of the visual hull [4][5]. Our algorithm to recovering real object shape is similar.

Registration. The most common approach to registering a virtual representation and the real object is to employ tracking systems. Devices, using magnetic fields, acoustic ranging, optical readings, retro-reflectors or gyros, are attached to the object and the sensor's reports are used to transform the virtual models. Hoffman attached a tracker to a real plate to register a virtual model of a plate that was rendered in a VE [6], and this allowed the participant to handle a real plate where the virtual plate appeared. Commercial products include the Immersion Corporation's

Cyberglove for hand tracking, and Measurand's ShapeTape, a flexible curvature-sensing device that reports its form.

Image-based algorithms, such Kanade's Virtualized Reality [7], IBVH, and this work, capture object motion by computing object representations anew from camera images.

Interactions. Collision detection between virtual objects is an area of *vast* research. Highly efficient and accurate packages, such as Swift++, detect collisions between polygonal objects, splines, and surfaces [8]. Hoff uses graphics-hardware accelerated functions to solve for collisions and generate penetration information [9]. Other work on collision detection between real and virtual objects first created geometric models of the rigid-body real objects and then used standard collision detection approaches [10].

Avatars and Interactions in VEs

The user is represented within the VE with a self-avatar, either from a library of self-avatar representations, a generic self-avatar, or no self-avatar. A survey of VE research shows the most common approach is a generic self-avatar—literally, one size fits all [1].

Avatars are typically represented with stylized virtual human models, such as those provided in commercial packages. Although these models contain a substantial amount of detail, they usually do not visually match a specific participant's appearance. Previous research hypothesizes that this misrepresentation of self is so detrimental to VE effectiveness, it will reduce how much a participant believed in the virtual world, his *sense-of-presence* [11]. Usoh concludes, "substantial potential presence gains can be had from tracking all limbs and customizing avatar appearance [12]."

Inputs to the VE are traditionally accomplished by translating hardware actions, such as button pushes or glove gestures, to actions such as grasping [13]. Commercial interaction devices include a tracked articulated glove with gesture recognition or buttons (Immersion's Cyberglove), tracked mouse (Ascension Technology's 6D Mouse), or tracked joystick with multiple buttons (Fakespace's NeoWand).

Studies have also been done on interaction devices, techniques, such as 3–D GUI widgets and physical interaction [14], and specifically engineering real objects for VE input, such as augmenting a doll's head with sliding rods [15].

REAL OBJECT RECONSTRUCTION

The reconstruction algorithm was originally presented at the ACM Symposium on Interactive 3D Graphics 2001 [16].

Introduction. We present a real-time algorithm for computing the visual hull of real objects that exploits the tremendous recent advances in graphics hardware. The visual hull technique examines only the silhouettes of the real objects, viewed from different locations. The projection of a silhouette image divides space into a volume that contains the real objects, and a remaining volume that does not. The intersection of the projections of silhouette images approximates the object shape [17]. Along with the IBVH work, this algorithm is one of the first for real-time object reconstruction.

Capturing Real Object Shape

The reconstruction algorithm, takes multiple, live, fixed-position video camera images, identifies newly introduced real objects in the scene (*image segmentation*) and then computes a novel view of the real objects' shape (volume-querying).

Image Segmentation. We assume that the scene will be made up of static background objects and foreground objects that we wish to reconstruct. The goal of this stage is to identify the foreground objects in the camera images of the scene.

We use the image segmentation technique of *image subtraction with thresholds* for extracting the objects of interest. When a live frame is captured, each pixel is compared against a reference image pixel of an "empty" scene (only the background). The pixels whose differences are greater than a threshold (to reduce the effect camera images noise) are labeled *object pixels* correspond to newly introduced objects. This is done for each camera at every frame and produces a set of object pixels ($S(O_i)$). The visual hull is the intersection of the projected right cones of the 2–D object pixels.

Volume-Querying. Given the object-pixels from image segmentation, we want to view the visual hull of the real objects. To do this, we use a method we call *volume-querying,* a variation on standard techniques for volume definition given boundary representations. Volume-querying asks, *given a 3–D point (P), is it within the visual hull (VH) of a real object in the scene?* P is within the visual hull iff for each camera i (with projection matrix M_C), P projects onto an object pixel.

$$P \ni \sim VH_{object} \text{ iff } \forall\, i \text{ such that } M_{C,\,i} * P \ni S(O_i). \tag{1}$$

To render the visual hull from a novel viewpoint, the view frustum volume is volume-queried. In effect, this asks which points in the view frustum are within the visual hull.

Accelerating Volume-Querying with Graphics Hardware. The graphics-hardware-accelerated functions of projected textures, alpha testing, and stencil testing in conjunction with the depth, stencil, and frame buffers are used for volume-querying.

After image segmentation, each camera's image, with the corresponding object-pixel data stored in the alpha channel (pixel alpha=1 for object pixels, else 0), is loaded into a texture. The camera image color values are not used in generating object shape.

For *P* to be within the visual hull, *P* must project onto an object pixel in each camera. Thus when rendering *P* with projected textures, *P* must be textured with an object pixel from each camera. *P* is rendered *n* times, and when rendering for the *i*th time, camera *i*'s texture is used, and the texture matrix is set to the $M_{C,i.}$ To apply a texel only if it is an object pixel, an alpha test to render texels with alpha = 1 is enabled.

A pixel's stencil buffer value is used to count the number of cameras that projected an object pixel onto *P*, and is initialized to 0. If *P* is textured by an object pixel from camera *i*, the pixel's stencil buffer is incremented. Once all *n* textures are projected, the stencil buffer will contain values [0, *n*]. A pixel is within the visual hull iff its stencil value is equal to *n*. The stencil buffer is then cleared of all pixels whose stencil value $< n$. The depth buffer value holds the distance of *P* from the novel viewpoint.

As the view frustum volume is continuous, we sample the volume with a set of planes perpendicular to the view direction, and completely filling the viewport. The planes are volume-queried from front to back. Each plane is rendered $n+1$ times, once with each camera's object-pixel map projected and once to clear pixels with a stencil value $< n$. These correspond to points on the plane that are within the visual hull. The frame and stencil buffers are not cleared between planes, and the depth buffer is the volume-sampled visual hull first visible surface from the novel viewpoint.

The number and spacing of the planes are user-defined. Given the resolution and location of the input cameras, we sample the volume with 1.5 cm spacing between planes for a meter in front of the user. By only volume-querying points within the view frustum, we only test elements that could contribute to the final image.

Capturing Real Object Appearance

Volume-querying only captures the real object shape. To capture the real object's appearance from the participant's point of view, a lipstick camera with a mirror attachment was mounted onto the HMD. Because of the geometry of the fixture, this camera had a virtual view that was essentially the same as the participant's. The image from this camera textures the visual hull. This particular camera choice finesses a set of difficult problems of computing the correct pixel color for the visual hull, which involves accounting for visibility and lighting. Figure 1 is a screenshot from the system.

This approach is not well suited for rendering other than from the participant's point of view. Coloring approaches are discussed in the original paper, but the results are far from satisfactory. The IBVH algorithm by Matusik computes the model and visibility and is a better for reconstruction from viewpoints other than the participant's [4][5].

Combining with Virtual Object Rendering

During the plane-sweeping step, the planes are rendered and volume-queried in the same coordinate system as used to render the VE. Therefore rendering the virtual objects into the same frame buffer and depth buffer correctly resolves occlusions between real objects and virtual objects. The real-object avatars are visually composited with the VE as shown in Figure 2. The real-object avatars can also be used lighting and shadowing between real and virtual objects, and were covered in the original paper.

Figure 1. Screenshot from our reconstruction system. Note the correct occlusion between the participant's hand (real) and the teapot handle (virtual).

Figure 2. The real-object avatars of the plate and user can interact (shadowing and collisions) with the virtual objects.

Analysis

Performance. The algorithm's overall work is the sum the image segmentation and volume-querying work. This analysis does not take into account the time and bandwidth of capturing new images, interprocessor communication, and VE rendering.

For each frame, the image segmentation work is composed of subtracting each camera image pixel from a background pixel, and comparing the result against a threshold value. Given n cameras with u x v resolution, u^*v^*n subtract and compares are required.

The volume-querying work has both a graphics transformation and a fill rate load. For n cameras, rendering l planes with u x v resolution and divided into an i x j camera-distortion correction grid, the geometry transformation work is $(2(n^*i^*j)+2)^*l$ triangles per frame. Volume-querying each plane computes u * v point volume-queries in parallel. Since every pixel is rendered n+1 times per plane, the *fill rate* = $(n+1)^*l^*u^*v$ per frame.

Accuracy. The final image of the visual hull is a combination of image segmentation, volume-querying, and visual hull sampling. How closely the final rendered image of the real-object avatar matches the actual real object has two separate components: how closely the shape matches, and how closely the appearance matches.

The primary source of error in shape between a real object and its real-object avatar is due to the visual hull approximation of the real object's shape. Fundamental to using the visual hull approaches, errors in real object shape approximation enforces a lower bounds of overall error, regardless of other sources of error [18].

Image segmentation errors (mislabeling a pixel as an object or background pixel) results from foreground objects similar in color to background objects, areas of high spatial frequency in the background, and changes in lighting. Segmentation errors incorrectly segment the volume (inside or outside the visual hull), but do not contribute to errors in the visual hull location.

The next error source is how closely the virtual volume that an object pixel sweeps out matches the physical space volume. This depends on the accuracy of the calculated intrinsic and extrinsic parameters. With a 1 m^3 reconstruction volume, camera rotation and resolution are the major factors that affect volume-querying accuracy. For example, 1° of rotational error results in 5.75 cm error in the reconstruction volume.

The camera resolution determines the minimum size of a foreground object to be reconstructed. The largest distance from a camera to a point in the reconstruction volume is 3.3 m. Using one field of the NTSC-resolution cameras (720x243) with 24° FOV lenses, a pixel sweeps out a pyramidal volume with at most a base of 0.58 cm by 0.25 cm.

The effect of camera calibration error on visual hull *location* is a bit more difficult to quantify, as this type of error would cause object pixels to sweep out a volume not registered with the physical space. It will also shift the projection of an object pixel, but this does not necessarily change the location of the visual hull.

The head tracker's noise, sub-millimeter in position and 0.1° in rotation, was not a significant component in reconstruction error. The primary factor that affects the rendering the visual hull is the spacing between the planes.

Our Experience: We attempt to reduce segmentation errors by draping dark cloth on surfaces to reduce high spatial frequency areas, keeping lighting constant and diffuse, and using foreground objects that were different in color from the background. Our Sony DFW-500 cameras had about a 2% color variation for the static cloth draped scene. The cameras are placed as close to the working volume as possible. There is an estimated 1 pixel of error for the rotation parameters and sub-millimeter error for the position parameters. We estimate 0.5 cm error for the center of the reconstruction volume is the lower bound for the certainty of the results for volume-querying a point. The plane spacing was 1.5 cm in the reconstruction volume.

The reconstructed shape is texture-mapped with the image from HMD mounted camera. The camera image was hand-tuned with interactive GUI sliders to keep the textured image registered to the real objects. We did not calibrate this front camera. We do not have an estimate for the error in visual hull appearance.

Other sources of error include: the lack of camera synchronization, system latency, and the variability in the position of the participant's head in the HMD. The most significant of these is the end-to-end system latency, estimated to be 0.3 seconds. The magnitude of the latency was such that participants recognized the lag and its effects on their ability to interact with virtual and real objects.

Implementation

Hardware. The reconstruction algorithm has been implemented in a system that reconstructs objects within a 5' x 4' x 3' volume above a tabletop. The system used three wall-mounted NTSC cameras (720x486 resolution) and one camera mounted on a Virtual Research V8 HMD (640x480 resolution). One camera was directly overhead, one camera to the left side of the table, and one at a diagonal about three feet above the table. Lab space and maintainability constrained the cameras' positions.

To compensate for the two fields, reconstruction was always done on the same field—field zero was arbitrarily chosen. While this increased the reconstruction error, latency was reduced and dynamic objects exhibited less shearing. The participant was tracked with the UNC HiBall, a scalable wide-area optical tracker mounted on the HMD.

The four cameras are connected to Digital In—Video Out (DIVO) boards on an SGI Reality Monster system. Whereas PC graphics cards could handle the transformation and pixel fill load of the algorithm, the SGI's video input capability, multiple processors, and high memory-to-texture bandwidth made it a better solution during initial development.

In implementation, the camera images contain non-linear distortions that the linear projected-texture hardware cannot process. Each plane is subdivided into a regular grid, and undistorted texture coordinates at the grid points are computed in software using a standard camera model. We have observed that dividing the plane into a 5 x 5 grid for undistorting the camera image improves visual hull shape accuracy.

In the past three years, other multiple camera algorithms have been implemented on a dedicated network of commodity PCs. With the increase of PC memory, bus, and device I/O bandwidth, a PC based system is now a viable solution and would also benefit from a short development cycle, speed upgrades, and additional features for new hardware.

We used five SGI graphics pipes: a *parent pipe* to render the VE and assemble the reconstruction results, a *video pipe* to capture video, two *reconstruction pipes* for volume-querying, and a *simulation pipe* to run simulation and collision detection. The reconstruction was done in a 320x240 window to reduce the fill rate requirements. The results were scaled to 640x480, which is the VE rendering resolution.

Performance. The implemented system runs on an SGI Reality Monster, and runs at 15–18 FPS for 1.5 cm spaced planes for 0.7 m deep (about 50 planes) in the novel view volume. The total work is $15.7 * 10^6$ subtracts and segmentation threshold tests per second, $0.23 * 10^6$ triangles per second are perspective-transformed, and the fill rate is $0.46 * 10^9$ per second. The latency is estimated at about 0.3 of a second.

The SGI can transform about $1.0 * 10^6$ triangles per second and has a fill rate of about $0.6 * 10^9$ pixels per second. For comparison, the current latest consumer graphics card, the nVidia GeForce4, can transform about $75.0 * 10^6$ triangles per second and has a fill rate of $1.2 * 10^9$ pixels per second. The fill rate requirements limits the number of planes with which we can sample the volume, which then limits reconstruction accuracy.

Accuracy Summary. For our setup, the overall total error in the visual hull shape is estimated at 0.5 cm and the rendering of the visual hull at 1.5 cm. One practical test we used was to move our hand with a finger (about 1 cm in diameter) extended around the reconstruction volume. We examined the reconstruction width of the finger to observationally evaluate error. The finger reconstruction was relatively constant throughout most of the working volume. This is inline with our estimates of 0.5 cm error for the visual hull shape, and 1.5 cm error for rendering the visual hull.

Advantages. The hardware-accelerated reconstruction algorithm benefits from the improvements in graphics hardware. It also permits using graphics hardware for detecting intersections between virtual models and the real-objects avatars (see later section).

The participant is free to extemporaneously bring in other real objects and naturally interact with the virtual system. For example, we implemented a VE with a virtual faucet and particle system. We observed participants cup their hands to catch the water, hold objects under the stream to watch particles flow down the sides, and comically try to drink the synthetic water. Unencumbered by additional trackers and intuitively interacting with the VE, participants exhibit uninhibited exploration.

Disadvantages. Sampling the volume with planes gives this problem $O(n^3)$ complexity. Large volumes would force a tradeoff between accuracy and performance.

Visibility and coloring, assigning the correct color to a pixel considering obscuration, is not handled well. This is not a problem since we are interested in a 1st person view and use an HMD-mounted camera for a high-resolution texture map. For novel viewpoint reconstruction, these are important issues to resolve.

COLLISION DETECTION

The *collision detection* and *collision response* algorithms, along with the lighting and shadowing rendering algorithms, enable the real objects to be dynamic inputs to simulations and provide a natural interface with the VE. That is, participants would interact with virtual objects the same way as if the environment were real.

For example, we will later show a participant parting a virtual curtain to look out a virtual window. The interaction between the real hand and virtual cloth involves first upon detecting the collision between hand and cloth, and then upon the cloth simulation's appropriately responding to the collision. Collision detection occurs first and computes information used by the application to compute the appropriate response.

The laws of physics resolve collisions between real objects. Standard collision detection packages handle collisions between virtual objects. We present an image-space algorithm to detect and allow the virtual objects to plausibly respond to collisions with real objects. We do not handle virtual objects affecting real objects due to collision.

Real Object Visual Hull—Virtual Model Collision Detection

Overview. Since the reconstruction algorithm does not generate a geometric model of the visual hull, we needed new algorithms to detect collisions between the real-object avatars and virtual objects. Similar to how the object reconstruction algorithm volume-queries the novel view frustum, the collision detection algorithm tests for collisions by volume-querying with the virtual objects primitives.

The *real-virtual* collision detection algorithm takes as inputs a set of n live camera images and virtual objects defined triangles. It outputs a set of points (CP_i) on the virtual object surface that are within a real-object avatar. It also estimates the following: point of first contact on the virtual object (CP_{obj}) and the visual hull (CP_{hull}), recovery vector (V_{rec}) and distance (D_{rec}), and surface normal at the point of visual hull contact (N_{hull}).

Assumptions. A set of simplifying assumptions makes interactive-time real-virtual collision detection a tractable problem.

1. *Only virtual objects can move or deform as a consequence of collision.*
2. *The real-object avatar and virtual object are considered stationary when resolving collisions.* With no real object motion information available, the algorithm cannot determine *how* or *when* the real and virtual objects came into collision. It simply suggests a way to move the virtual object out of collision.
3. *There is at most one collision between a virtual and real-object avatar at a time.* For multiple intersections, we heuristically choose one as the point of contact.
4. *The real objects that contribute to the visual hull are treated as a single object.* The system cannot distinguish between the real objects that form a visual hull.
5. *Collisions are detected relatively shortly after a virtual object enters the visual hull, and not as the virtual object is exiting the visual hull.* This assumes the simulation time step (frame rate) is fast compared to the dynamics of the objects.

Detecting Collisions. Collision points, $CP_{i,}$ are points on the surface of the virtual object that are within the visual hull. As the virtual surfaces are continuous, the set of collision points is a sampling of the virtual object surface.

The real-virtual object collision detection algorithm uses volume-querying. In novel viewpoint reconstruction, we volume-queried points in the view frustum volume to determine which were inside the visual hull. Collision detection volume-queries with the triangles defining the virtual object's surface to determine if any parts of the surface is inside the visual hull. If any part of a triangle lies within the visual hull, the object is intersecting a real-object avatar, and a collision has occurred.

Everything else (the textures, stencil testing, etc.) is the same. A collisions occurs if any pixel has a stencil buffer = n, thus indicated some part of a triangle, and in turn a virtual object, is within the visual hull. If the triangle is projected 'on edge' during volume-querying, the sampling of the triangle surface during scan-conversion will be sparse, and collision points could be missed. No one viewpoint will be optimal for all triangles. Thus, each triangle is volume-queried in its own viewport, such that the triangle's projection maximally fills the viewport.

After all the triangles are volume-queried, the frame buffer is read back. The collision pixels (stencil buffer value = n) are unprojecting from screen space coordinates (u, v, depth) to world space coordinates (x, y, z) to produce the 3–D coordinates of a collision point. These 3–D points form a set of collision points, $CP_{i,}$ for that virtual object. As an optimization, collision detection is first done with virtual object bounding boxes, and if there are collisions, on a per-triangle test is done.

How a simulation utilizes this information is application- and even object-dependent. This division of labor is similar to current collision detection algorithms [8]. We provide tools to move the virtual object out of collision with the real object.

Recovery from Interpenetration. We present one approach to use the collision information to generate a plausible response. The first step is to move the virtual object out of collision. We estimate the point of first contact on the virtual object, $CP_{obj,}$ with the collision point farthest from the virtual object's reference point, $RP_{obj.}$ The default RP_{obj} is the center of the virtual object.

Our estimate to move the virtual object out of collision by the shortest distance is along a recovery vector, V_{rec} defined as the vector from CP_{obj} to $RP_{obj.}$ This vector works well for most objects; though the simulation can specify V_{rec} for virtual objects with constrained motion, such as a hinged door, for better object-specific results. V_{rec} crosses the visual hull boundary at the *hull collision point*, $CP_{hull.}$ CP_{hull} is an estimate of the point of contact on the visual hull, and to where CP_{obj} will be backed out.

To find $CP_{hull,}$ V_{rec} is searched from RP_{obj} towards CP_{obj} for the first point within the visual hull. This is done by volume-querying an isosceles triangle ABC, $A = CP_{obj}$ and the base, BC, is bisected by $V_{rec.}$ Angle BAC (at CP_{obj}) is set small (10°) so that AB and AC intersects the visual hull near $CP_{hull,}$ and the height is set relatively large (5 cm) so the triangle base is likely to be outside the visual hull. ABC is volume-queried in the entire window's viewport, and from a viewpoint along the triangle normal and such that V_{rec} lies along a scan line. CP_{hull} is found stepping along the V_{rec} scan line for the first pixel within the visual hull. Unprojecting the pixel from screen to world space yields $CP_{hull.}$

The *recovery distance*, $D_{rec,}$ is the distance between CP_{obj} and $CP_{hull,}$ and is the distance along V_{rec} required to move CP_{obj} outside the visual hull. It is not necessarily the *minimum separation distance*, as is found in some collision detection packages [8][9].

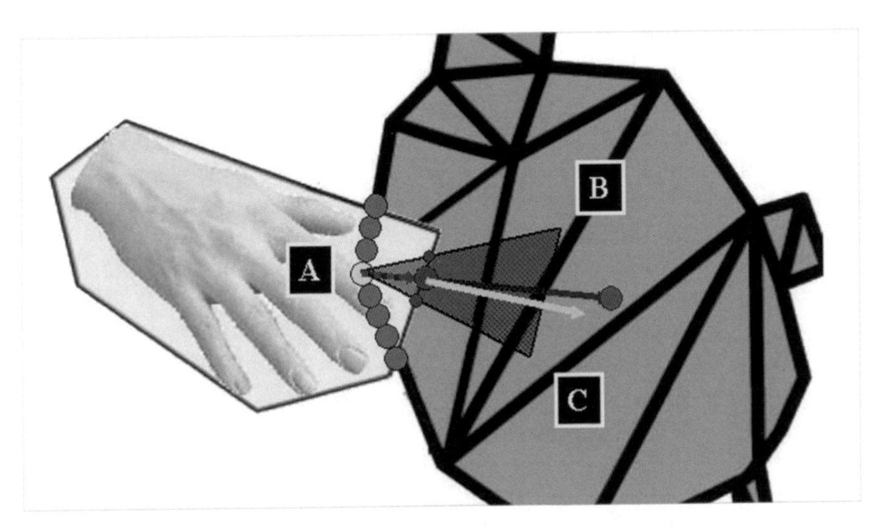

Figure 3. Collision detection returns CP_i (points of hand within teapot) and CP_{obj} (A). Collision response returns V_{rec} (vector towards center of teapot), CP_{hull} (where the vector intersects the visual hull boundary), D_{rec} (distance between A and CP_{hull}), and N_{hull} (the normal at CP_{hull}).

To compute the visual hull collision point surface normal, N_{hull}, we locate 4 points on the visual hull surface near CP_{hull}. Stepping along *BA* and *CA* and finding where they intersect the visual hull boundary determines points *I* and *J*. A second triangle, *DAE*, is constructed that is similar to *ABC* and lies in a plane roughly perpendicular to *ABC*. Points *K* and *L* are located by stepping along *DA* and *EA*. N_{hull} is the cross product of *IJ* and *KL*.

Analysis

Performance. Given *n* cameras, virtual objects with *m* triangles, and a $u \times v$ viewport in a $x \times y$ window: the geometry transformation cost is $(n * m)$, fill rate cost is $(n*m*u*v)/2$, and $(x*y)$ pixel readbacks and compares per frame. Our window and curtain hybrid environment had 720 triangles that made up the curtains. We used 10 x 10 viewports in a 400 x 400 window for collision detection, which ran at 6 FPS. The collision detection work was 13,000 triangles transformed per second, and 648,000 pixels per second fill rate, and 160,000 pixel readbacks and compares. The collision response work was 2 triangles and 480,000 pixels fill rate per virtual object in collision. As this was a first implementation, there are many optimizations that should improve the performance.

Accuracy. The collision detection accuracy depends on image segmentation, camera models (previously discussed), and the volume-querying viewport. The size of the viewport is proportional to the volume-querying spatial sampling accuracy, and inversely proportional to the speed and number of triangles that can be queried in a single pass. The accuracy of collision detection for a u x v resolution viewport (if $u = v$, viewport layout is easier) and a triangle with x x y bounding box (in world space) is x/u by y/u.

The size of virtual object triangles will vary, but typical tabletop objects had triangles less than 2 cm, which would have 0.2 cm x 0.2 cm collision point detection error. For example in the cloth system had a collision detection resolution of 0.75 cm x 0.3 cm.

For collision response, the accuracy of the CP_{hull} point impacts D_{rec} and $N_{hull.}$ The error in finding CP_{hull} along the V_{rec} is the length of triangle ABC's major axis divided by the horizontal length of collision response window (assuming a square window). With a 400 x 400 rendering window, this results in .0125 cm error for detecting $CP_{hull.}$ The accuracy of $N_{hull,}$ depends on the surface topology (effected by camera resolution), the distance from these points to $CP_{hull,}$ and the distance from CP_{hull} to $CP_{obj.}$ The magnitude of these errors is smaller than the error in the visual hull location and visual hull shape.

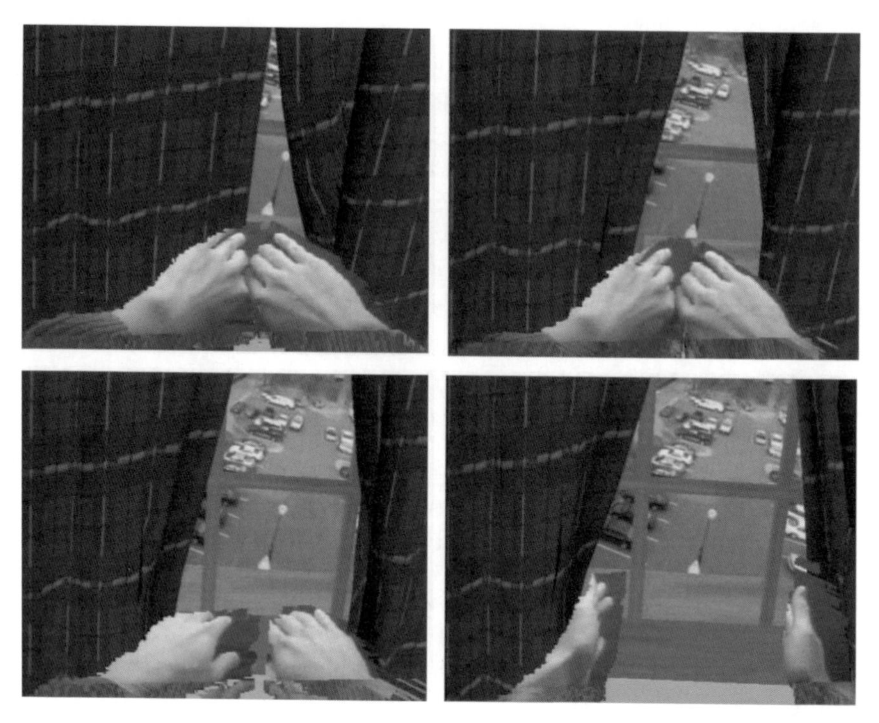

Figure 4. Sequence of images with the user naturally parting virtual curtains to look out a virtual window.

Implementation. Figure 4 is a sequence of frames of a user pushing aside a curtain with his hands. This shows the use of the algorithm with a deformable virtual object with constrained motions (specified V_{rec} to the collision response algorithm). Now, when trying to move the cloth nodes out of collision, the motion is primarily in the constrained vector direction. We have also prototyped particle systems, lighting, and shadowing simulations interacting with the real-object avatars.

USER STUDY

Motivation. Two components that the implemented hybrid environment provides are interacting with real objects and visually faithful self-avatars. But does that even benefit VE tasks? We conducted a study to identify the effects of interaction methodologies and avatar visual fidelity on task performance and *sense-of-presence* while conducting a cognitive manual task. Compared to virtual objects and generic self-avatars,

- Does interacting with real objects improve task performance?
- Does seeing a visually faithful self-avatar improve sense-of-presence?

Experiment

Design Decisions. In devising the task, we sought to abstract tasks common to VE design applications. Through surveying production VEs [1], we noted that a substantial number of VE goals involve participants doing spatial cognitive manual tasks. For example, in layout applications, users evaluate different configurations and designs.

The task we designed is similar to, and based on, the block-design portion of the Wechsler Adult Intelligence Scale (WAIS). Developed in 1939, the WAIS is a test widely used to measure intellectual quotient, IQ [19]. The block-design portion measures reasoning, problem solving, and spatial visualization.

Description. Participants manipulated a number of identical 3″wooden blocks to make the top face of the blocks match a target pattern. The faces represented the possible quadrant-divided white-blue patterns. There were two sizes of target patterns, *small* four block patterns in a 2x2 arrangement, and *large* nine block patterns in a 3x3 arrangement.

Design. The user study was a between-subjects design. Each participant performed the task in a real space environment (RSE), and then in one of three VE conditions. The independent variables were the interaction modality (real or virtual blocks) and the self–avatar fidelity (generic or visually faithful). The three VE conditions were:

- Virtual objects with generic self-avatar (purely virtual environment)
- Real objects with generic self-avatar (hybrid environment)
- Real objects with visually faithful self-avatar (vis-faithful hybrid environment)

Experiment Conditions. *Real Space Environment (RSE)*—participants manipulated nine wooden blocks inside a rectangular 36" x 25" x 18" enclosure. *Purely Virtual Environment (PVE)*—participants wore Fakespace Pinchgloves, each tracked with Polhemus Fastrak trackers, and a Virtual Research V8 HMD. Using pinching gestures, the participant manipulated virtual blocks with his self-avatar. Hybrid Environment (HE)—participants wore yellow dishwashing gloves and the HMD. Within the VE, participants handled physical blocks, identical the RSE blocks, and saw a self-avatar with accurate shape and generic appearance (dishwashing gloves). *Visually-Faithful Hybrid Environment (VFHE)*—similar to the HE except the participants did not wear gloves. The self-avatar was visually faithful as the shape reconstruction was textured with images from a HMD mounted camera. The participant saw an image of himself, warts and all.

Virtual Environment. The VE was identical in all three of the virtual conditions (PVE, HE, VFHE). The room had several virtual objects, including a lamp, a plant, and a painting, along with a virtual table that was registered with a real Styrofoam table. The enclosure in the RSE was also rendered in the VE, but was rendered with transparency.

All the VE conditions were rendered on an SGI Reality Monster. The PVE ran on one rendering pipe with four raster managers at a minimum of 20 FPS. The HE and VFHE ran on four rendering pipes at a minimum of 20 FPS for virtual objects, and 12 FPS for reconstructing real objects. The participant wore a Virtual Research V8 HMD (640x480 resolution in both eyes) that was tracked with the UNC HiBall tracking system.

We expect a participant's RSE (no VE equipment) performance would produce the best results, as the interaction and visually fidelity were optimal. We compared how closely a participant's task performance in VE was to their RSE task performance. We compared the reported sense-of-presence in the VEs to each other. The PVE is a plausible approach with current technology. The HE evaluates the effect of real objects on task performance and presence. The VFHE adds visually faithful self-avatars.

Measures. For *task performance* we measured the time (in seconds) for a participant to arrange the blocks to exactly match the target pattern. The dependent variable was the difference in a participant's task performance between the RSE condition and VE condition. For *sense-of-presence,* the dependent variable was the sense-of-presence scores from the Steed-Usoh-Slater Presence Questionnaire (SUS) [20].

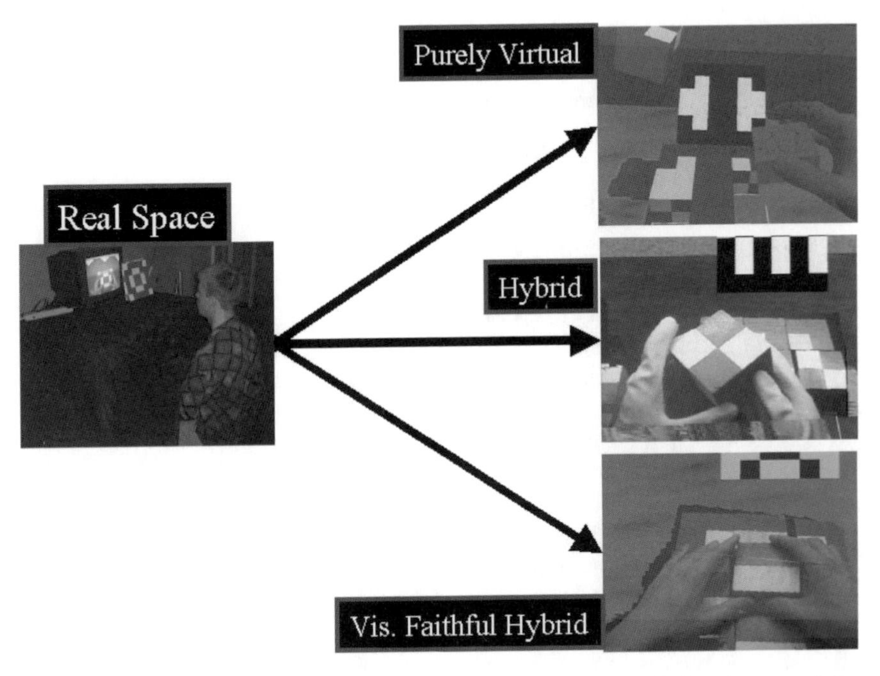

Figure 5. Each participant performed the task in the RSE and then in one of the three VEs.

Finally, we conducted a debriefing interview and administered the Guilford-Zimmerman Aptitude Survey, Part 5: Spatial Orientation and the Kennedy–Lane Simulator Sickness Questionnaire.

Experimental Procedure. After completing the initial forms and questionnaires, the participant did the task in the RSE. Participants performed a series of practice patterns, three small and then three large. Next, the participant completed six timed patterns, three small and three large. Then, the participant conducted the task in a VE condition. Following a period of adaptation to the VE, the participant practiced on two small and two large patterns, and then did two small and two large timed patterns. Finally, the participants were interviewed about their impressions of and reactions to the session, and completed the final series of questionnaires.

Hypotheses. *Task Performance*: Participants who manipulate real objects in the VE (HE, VFHE) will complete the spatial cognitive manual task significantly closer to their RSE task performance than will participants who manipulate virtual objects (PVE).

Sense-of-presence: Participants represented in the VE by a visually faithful self-avatar (VFHE) will report a higher *sense-of-presence* than will participants represented by a generic self-avatar (PVE, HE).

We expect interacting with real objects improves task performance regardless of self-avatar visual fidelity and generic self-avatars would have similar effects on presence regardless whether there were real or virtual objects.

Results and Discussion

Forty participants completed the study, thirteen each in the purely virtual environment (PVE) and hybrid environment (HE), and fourteen in the visually-faithful hybrid environment (VFHE). The participants were primarily male (thirty-three) UNC undergraduate students (thirty-one). They reported little prior VE experience (M=1.37), high computer usage (M=6.39), and moderate—1 to 5 hours a week—computer/video game play (M=2.85), on [1..7] scales. There were no significant differences between the groups. We use a two-tailed t-test with unequal variances and an =0.05 level.

Task Performance. For small and large patterns, both VFHE and HE task performances were significantly better than PVE task performance. The difference in task performance between the HE and VFHE was not significant at the =0.05 level. As expected, performing the block-pattern task took longer in any VE than it did in the RSE: The PVE participants took about three times as long to do the task as they did in the RSE. The HE and VFHE only took about twice as long (Table 1).

For the case we investigated, *interacting with real objects provided a quite substantial performance improvement over interacting with virtual objects for cognitive manual tasks.* Although task performance in all the VE conditions was substantially worse than in the RSE, the task performance of HE and VFHE participants was significantly better than for PVE participants. There

Table 1. Task Performance [seconds] and SUS Sense of Presence [0...6]. Mean (Standard Deviation).

	RSE (*n*=40)	PVE (*n*=13)	HE (*n*=13)	VFHE (*n*=14)
Small Patterns	16.81 (6.34)	47.24 (47.24)	31.68 (5.65)	28.88 (7.64)
Large Patterns	37.24 (8.99)	116.99 (32.25)	86.83 (26.80)	72.31 (16.41)
SUS Score		3.21 (2.19)	1.86 (2.17)	2.36 (1.94)

Table 2. Between Groups Task Performance and Presence (t-test w/ unequal variances, p-value).

	PVE-RSE vs VFHE-RSE	PVE-RSE vs HE-RSE	VFHE-RSE vs HE-RSE
Small Patterns	(t = 3.32, 0.0026)**	(t = 2.81, 0.0094)**	(t = 1.02, 0.32)
Large Patterns	(t = 4.39, 0.00016)***	(t = 2.45, 0.021)*	(t = 2.01, 0.055)+
	(t = 1.10, 0.28)	(t = 1.10, 0.28)	(t = 1.10, 0.28)

**significant at =0.05, **significant at =0.01, ***significant at =0.001, +requires further investigation.*

is a slight difference between HE and VFHE performance (Table 2, p=0.055). We do not have a hypothesis for this result.

Sense-of-presence. We augmented the standard Steed-Usoh-Slater Presence Questionnaire with questions that focused on the participants' perception of their avatars. Although interviews showed visually faithful self-avatars (VFHE) were preferred, *there was no statistically significant difference in reported sense-of-presence compared to those presented a generic self-avatar (HE and PVE).*

There were no statistically significant differences at the =0.05 level between any of the conditions for any of the sense-of-presence questions. Slater and Usoh cautions against the use of the SUS Questionnaire to compare presence across VE conditions, but also points out that no current questionnaire supports such comparisons [11].

Other Factors. Simulator sickness and spatial ability were not significantly different between the groups at the =0.05 level. Spatial ability was moderately correlated (r=–0.31 for small patterns, and r=–0.38 for large patterns) with performance.

Participant Interviews. Participants in all groups responded that the self-avatar, block task, head tracking, and virtual objects improved their sense-of-presence.

We noticed a trend in comments on self-avatar realism. All PVE and HE participant comments related to motion accuracy, "Everything I did with my hands, it followed." All VFHE participant comments related to visual accuracy, "[It was] just the same as in reality… I didn't even notice my hands." We hypothesize *kinematic fidelity of the self-avatar is more important than visual fidelity for sense-of-presence.*

Most HE and VFHE participants noted the reconstruction system's noise and lag. Some of the HE and VFHE participants noted that the tactile feedback of handling real objects increased their sense of presence, while 43% of the PVE participants felt that the virtual blocks reduced their sense-of-presence. Almost all the participants (93%) of the PVE felt the interaction was unnatural, compared to only 13% in the HE. VFHE participants became comfortable interacting with the VE significantly more quickly (1.50 to 2.36 practice patterns) than PVE participants ($T_{26} = 2.83$, p=0.0044).

Interesting Results. Rotating the block, followed by selection and placement of blocks, dominated the difference in times between VE conditions. Both were improved through the natural interaction, motion constrains, and tactile feedback of real blocks.

Recall that while the PVE participant made a pinching gesture to pick up a block, visually they saw the avatar hand grasp a virtual block. This misregistration caused 25% of the participants to forget the pinching mnemonic and try a grasping action (which at times did not register with the pinch gloves).

Conclusions. Interacting with real objects significantly improves task performance over interacting with virtual objects in spatial cognitive tasks, and more importantly, it brings performance measures closer to that of doing the task in real space. *Handling real objects makes task performance and interaction in the VE more like the actual task.*

Motion fidelity is more important than visual fidelity for self-avatar believability. We believe that a visually faithful self-avatar is better than a generic self-avatar, but from a sense-of-presence standpoint, the advantages do not seem very strong.

CASE STUDY: NASA COLLABORATION

Driving Problems. To evaluate the potential of this technology in a real world task, we applied our prototype to an assembly verification task, we have begun collaborating with the NASA Langley Research Center (LaRC).

NASA LaRC payload designers are interested in examining models of payload subsystems for assembly verification and assembly training. They want to discern possible assembly, integration, and testing problems early in the project development cycle. Since different subsystems are separately subcontracted out, the integration stage always generates compatibility and layout issues. Layout issues result in schedule delays, equipment redesign, or makeshift engineering fixes.

Early in development, the payload designs are stored as CAD models, and the assembly procedure is a step-by-step instruction list. Later in development, simplified physical mock-ups are manufactured for design verification and layout. We believe hybrid VEs can be an effective tool between these stages to enable designers to test configurations using the final assembly personnel, real tools and parts.

Payload Spacing Experiment. We received CAD models of a photon multiplier tube (PMT), a weather imaging satellite subsystem, currently under development. Next, we abstracted a task similar to common assembly steps, such as attaching components and cables to connectors. The PMT model, along with two other payloads (payloads A and B), was rendered in the VE. The hybrid system performed collision detection between the virtual payloads and the real-object avatars. The task was to screw a cylindrical shield (mocked-up as a PVC pipe) into a receptacle and then plug a power connector into an outlet inside the shield (Figure 6). The task was to determine how much space was required between the *top* of the PMT box and the *bottom* of payload A.

Experimental Procedure. Four NASA LaRC payload designers participated in the case study. Before attempting the task, we provided task information in approximately the same manner as they receive it in actual de-

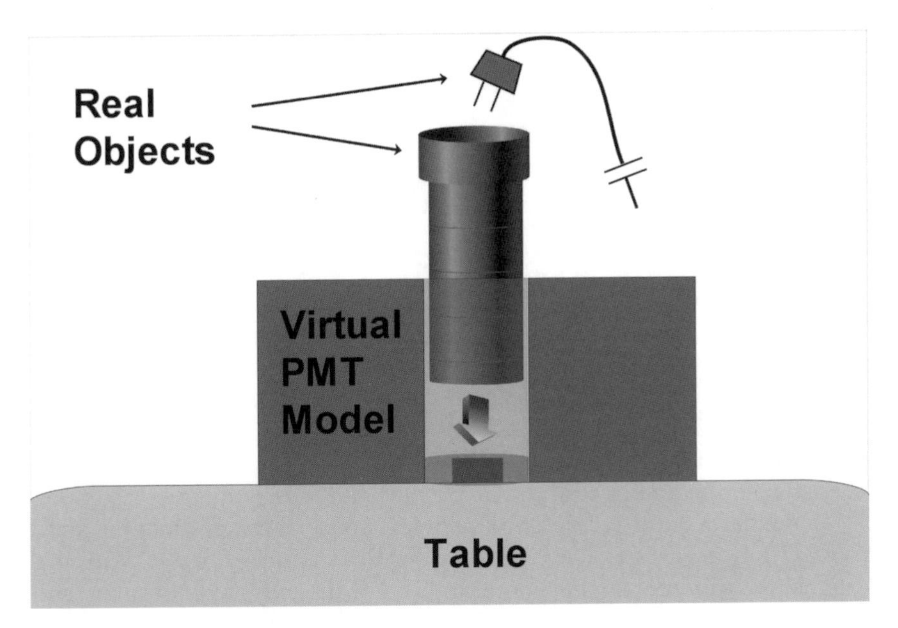

Figure 6. Task Objective: Determine space required between the PMT and payload A to perform the shield and cable-fitting task.

Figure 7. What the participant saw. The real objects (pipe and cable) among virtual objects (table and payload models).

Table 3. LaRC Participant Responses and Task Results (cm)

Between Payload A and the PMT,	#1	#2	#3	#4
(Pre) How much space is necessary?	14.0	14.2	15.0–16.0	15.0
(Pre) How much space would you actually allocate?	21.0	16.0	20.0	15.0
Actual space required (determined in VE)	15.0	22.5	22.3	23.0
(Post) How much space would you actually allocate?	18.0	16.0 (alter tool)	25.0	23.0

sign evaluation and surveyed them on the space needed between the PMT and payload A. Each participant then performed the pipe insertion and power cable attachment procedure in the hybrid system. After a period of VE adjustment, participants picked up the pipe and eased it into the center cylindrical assembly while trying to avoid colliding with any of the virtual payloads. After the pipe was lowered into the cylindrical shaft of the PMT, they snaked the power cord down the tube and inserted it into the outlet.

As the participant asked for more or less space, the experimenter could dynamically adjust the space between the PMT and payload A. With this interaction, different spatial configurations of the two payload subassemblies could be quickly evaluated.

Results. Given that the pipe had a length of 14 cm and a diameter of 4 cm:

Participant #1 was able to complete the task without using any tool, as the power cable was stiff enough to force into the outlet. Since an aim was to impress upon the participants the possibility of requiring unforeseen tools in assembly or repair, we used a more flexible cable for the remaining participants. While trying to insert the power cable, participants #2, 3, and 4 asked for a tool to assist. They were handed a set of tongs and were then able to complete the power cable insertion task. This required increasing the spacing between the PMT and Payload A from 14 cm to an average of 24 cm.

Whereas in retrospect it was obvious that the task would not be easily completed without a tool, none of the designers anticipated this requirement. We believe the way the assembly information was provided (diagrams, assembly documents and drawings), made it difficult for designers, even though each had substantial payload development experience, to catch subtle assembly integration issues. On average, the participants allocated 5.6 cm too little space between the payloads on their pre-experience surveys.

The hybrid VE system provided identifiable benefits over purely virtual approaches. Participants interacted with virtual objects as if they were real, including avoiding contact of the payload. "You just don't touch flight hardware". Accommodating tools extemporaneously, without additional modeling or development, enabled easy evaluation of multiple layouts, approaches, and tools. The pipe threads and cable socket provided important motion constraints that aided in interacting with these objects.

Debriefing. The participants were *extremely* surprised that both a tool and substantial additional space were required. The participants commented that the financial cost of the spacing error could range from moderate (keeping personnel waiting until a design fix was implemented) to extreme (launch delays). Time is the most precious commodity in payload development, and identifying the spacing error would save days to weeks. NASA LaRC payload designers remarked that VEs and object reconstruction VEs would be useful for assembly training, hardware layout, and design evaluation.

SUMMARY

This work investigated the methods, usefulness, and application of a hybrid environment capable of incorporating dynamic real objects. We developed real-time algorithms for generating virtual representations of real objects and collision management algorithms to handle interactions between real and virtual objects

We evaluated the algorithms from an engineering, theoretical, usability, and application standpoint. We conducted studies to examine the effects of interaction modality and avatar fidelity on task performance and sense-of-presence. We found that interacting with real objects significantly improves task performance for spatial cognitive tasks. We did not find the avatar visual fidelity affected sense-of-presence. We have begun applying our system to a NASA LaRC an assembly verification task. Initial trials show promise on the applicability of hybrid environments to aid in payload development.

ACKNOWLEDGEMENTS

I would like to thank Dr. Frederick P. Brooks, Jr. for his support and guidance. I want to acknowledge Samir Naik and Professor Mary Whitton for their invaluable contribution to this research. Finally, I would like to thank the Link Foundation and the University of North Carolina at Chapel Hill for the financial support.

REFERENCES

[1] F. Brooks. "What's Real About Virtual Reality?" *IEEE Comp. Gr. and App.*, Vol 19, No. 6, pp. 16–27. (1999)

[2] Sutherland, I. "The Ultimate Display", In *Proceedings of IFIP* 65, Vol 2, pp 506. (1965)

[3] K. Daniilidis, J. Mulligan, R. McKendall, G. Kamberova, D. Schmid, and R. Bajcsy. "Real-Time 3–D Tele-immersion", In *The Confluence of Vision and Graphics,* A Leonardis *et al.* (Eds.), Kluwer Academic Publishers, pp. 253–266. (2000)
[4] W. Matusik, C. Buehler, R. Raskar, S. Gortler and L. McMillan. "Image-Based Visual Hulls", In *Proceedings of ACM SIGGRAPH 2000,* Annual Conference Series, pp. 369–374. (2000)
[5] W. Matusik, C. Buehler, and L. McMillan. "Polyhedral Visual Hulls for Real-Time Rendering", In *Proceedings of 12th Eurographics Workshop on Rendering,* London, England, pp. 115–125. (2001)
[6] H. Hoffman. "Physically Touching Virtual Objects Using Tactile Augmentation Enhances the Realism of Virtual Environments", In *Proceedings of the IEEE Virtual Reality Annual International Symposium '98,* Atlanta GA, p. 59–63. IEEE Computer Society, Los Alamitos, California. (1998)
[7] S. Baba, H. Saito, S. Vedula, K.M. Cheung, and T. Kanade. "Appearance-Based Virtual-View Generation for Fly Through in a Real Dynamic Scene", In *Proceedings of VisSym '00* (Joint Eurographics—IEEE TCVG Symposium on Visualization). (2000)
[8] S. Ehmann and M. Lin. "Accurate Proximity Queries between Polyhedra Using Surface Decomposition", *Computer Graphics Forum (Proceedings of Eurographics).* (2001)
[9] K. Hoff, A. Zaferakis, M. Lin, and D. Manocha. "Fast and Simple 2–D Geometric Proximity Queries Using Graphics Hardware", *2001 ACM Symposium on Interactive 3–D Graphics,* pp. 145–148. (2001)
[10] D. Breen, E. Rose, R. Whitaker. "Interactive Occlusion and Collision of Real and Virtual Objects in Augmented Reality", Munich, Germany, European Computer Industry Research Center. (1995)
[11] M. Slater and M. Usoh. "The Influence of a Virtual Body on Presence in Immersive Virtual Environments", *VR 93, VR International, Proceedings of the Third Annual Conference on Virtual Reality,* London, Meckler, pp 34–42. (1993)
[12] M. Usoh, K. Arthur, *et al.* "Walking > Virtual Walking > Flying, in Virtual Environments", *Proceedings of SIGGRAPH 99,* pp. 359–364, Computer Graphics Annual Conference Series. (1999)
[13] D. Bowman and L. Hodges. "An Evaluation of Techniques for Grabbing and Manipulating Remote Objects in Immersive Virtual Environments", In *Proceedings 1997ACM Symposium on Interactive 3–D Graphics.* eds. by M. Cohen and D. Zeltzer, pp. 35–38. ISBN 0–89791–884–3. (1997)
[14] R. Lindeman, J. Sibert, and J. Hahn. "Hand-Held Windows: Towards Effective 2D Interaction in Immersive Virtual Environments", In *IEEE Virtual Reality.* (1999)
[15] K. Hinckley, R. Pausch, J. Goble, and N. Kassell. "Passive Real-World Interface Props for Neurosurgical Visualization", In *Proceedings of the 1994 SIG-CHI Conference,* pp 452–458. (1994)
[16] B. Lok. "Online Model Reconstruction for Interactive Virtual Environments", In *Proceedings 2001 ACM Symposium on Interactive 3–D Graphics,* Chapel Hill, N.C., 18–21, pp. 69–72, 248. (2001)
[17] A. Laurentini. "The Visual Hull Concept for Silhouette-Based Image Understanding", *IEEE Transactions on Pattern Analysis and Machine Intelligence,* Vol 16, No. 2, 150–162. (1994)

[18] W. Niem. *"Error Analysis for Silhouette-Based 3D Shape Estimation from Multiple Views", Proceedings on International Workshop on Synthetic—Natural Hybrid Coding and Three Dimensional Imaging* (IWSNHC3DI'97), Rhodos. (1997)
[19] D. Wechsler. *The Measurement of Adult Intelligence, 1st Ed.,* Baltimore: Waverly Press Inc. (1939)
[20] M. Usoh, E. Catena, S. Arman, and M. Slater. "Using Presence Questionnaires in Reality", *Presence: Teleoperators and Virtual Environments,* Vol. 9, No. 5, pp. 497–503. (2000)

PART III

OCEAN ENGINEERING AND INSTRUMENTATION

Measuring Pore Pressure in Marine Sediments with Penetrometers: Comparison of the Piezoprobe and DVTP-P Tools in ODP Leg 204

Brandon Dugan

Department of Geosciences

Penn State University, University Park PA 16801

Research Advisor: Dr. Peter B. Flemings

ABSTRACT

Fugro-McClelland Marine Geosciences Inc.'s piezoprobe, a penetration-based tool used to determine pore pressure and hydrologic properties within a borehole, was deployed for the first time in the Ocean Drilling Program (ODP) on ODP Leg 204 in July 2002. Analysis of the piezoprobe data suggests that *in situ* pore pressure is 9.5 MPa, which is approximately the hydrostatic pressure (9.53 MPa). The piezoprobe deployment and modeling of the results provides one of the first measurements of *in situ* permeability made within the borehole. From the piezoprobe dissipation data, we estimate *in situ* permeability of approximately 1.5×10^{-17} m^2 for the hemipelagic clay. This is consistent with laboratory-measured permeability (~$1 \text{x} 10^{-17}$ m^2) on hemipelagic clay samples from nearby ODP Site 892. The piezoprobe results were compared to a Davis-Villinger Temperature/Pressure Probe (DVTP-P) measurement made at the same depth, and in the same lithology, but in an adjacent borehole. The DVTP- P is also a penetration-based tool, however it has a much wider probe diameter. The DVTP-P generated a higher peak pressure that did not dissipate as much as the piezoprobe pressure, which resulted in a DVTP-P estimate of *in situ* pressure that nearly equals the overburden stress. The results suggest that a narrow diameter probe like the piezoprobe can be used to rapidly determine *in situ* pressure and hydrologic properties in sites investigated by the Ocean Drilling Program.

INTRODUCTION

Rock deformation, sediment strength, and regional fluid fluxes are directly related to pore pressure and hydrologic properties of the sub-seafloor sediments. Three geologic systems that can be more completely defined through direct measurements of pressure are (1) fluid flow and stability of continental margins, (2) fault activation, decollement location and propagation, and geometry in accretionary prisms, and (3) free gas and water migration, hydrate formation, and rock strength in gas hydrate provinces.

The role that pore fluids have in sculpting continental slope geomorphology has intrigued scientists since the diverse structure of slopes was identified [1, 2] (Figure 1A). Excess fluid pressure has been attributed to landslides and failures on low angle slopes that would not fail without excess pressure [3, 4]. More recently focused fluid migration along permeable layers has been invoked as a major contributor to the timing and distribution of sediment deformation and failure [5–8]. Models predict the magnitude of pressure required to generate slope instability and provide insights into the origins of the required excess pressure. Relatively few direct measurements exist to test the models, so the models are typically compared to pressure estimates from proxy data such as porosity [9] or seismic velocity [10].

Fluid migration within accretionary complexes has been described for its importance to heat and chemical transport [11] (Figure 1B). Fluids have also been cited as a driving force in the geometry and structure of accretionary complexes [12–14]. Porosity and seismic data have been used with models to estimate pressure, flow paths, and fluid fluxes [15, 16, 17]. These models also constrain the contribution of fluids to deformation, chemical transport, and heat flow. Validation of these models and their interpretations has not been extensive because of the lack of direct pressure measurements. The Ocean Drilling Program (ODP) has started to collect direct measurements of pressure, temperature, and pore fluid chemistry with long-term observatories (Circulation Obviation Retrofit Kit [CORKs], Advanced Circulation Obviation Retrofit Kit [ACORKs]) [18–20].

The pressure and stress in gas hydrate provinces is not well defined, is lacking robust multiphase models, and has very few direct observations. These pressures and stresses, however, are critical to the dynamics of this multiphase system. Fluid pressure impacts the solubility of gas in water, governs the stability of gas hydrate [21, 22], defines the permeability of the system, and pressure gradients dictate the flow field (Figure 1C). A detailed analysis of the complex hydrate system is required: (1) to define the volume of gas stored as hydrate and as free gas beneath hydrate [21]; (2) to understand the mechanics by which gas migrates and is released [23–25]; (3) to characterize the role of

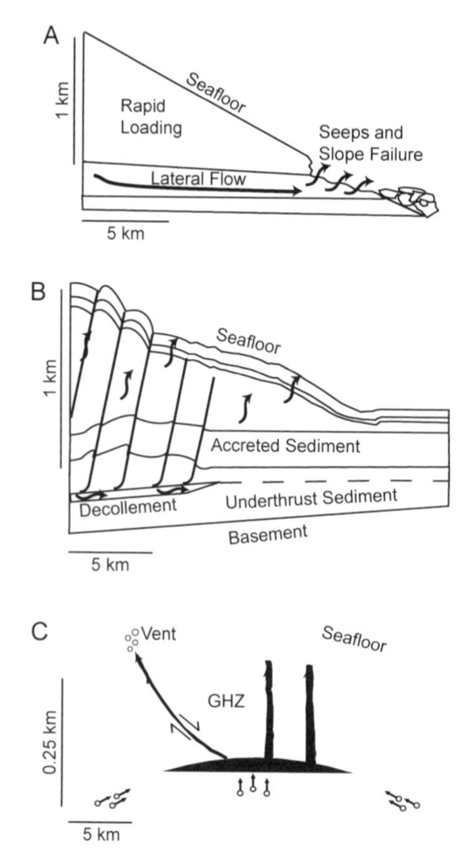

Figure 1. Direct pressure observations are necessary to describe a variety of sub-seafloor processes and seafloor geomorphology. Arrows illustrate flow paths that have been postulated for the systems, but require direct measurements to verify. (A) Continental slopes are environments where slope failure and seeps are common. High fluid pressures are often attributed to failure along low angle slopes but few direct measurements of *in situ* pressure have been collected to test the models. (B) Pore fluid pressure affects the flow of fluids along the decollement and within faults in accretionary complexes. Pressures also control the geometry of the accretionary complex, e.g. the angle between the decollement and seafloor is small when excess pressures are high and is large when pressures are hydrostatic. The transition from the proto-decollement (minimal to no deformation) to the decollement (failure and faulting) is believed to be a function of flow and fluid pressure. (C) Gas hydrate provinces are dynamic hydrologic systems where gas and water pressures affect the formation and dissociation of gas hydrate. Permeability and gas storage are interpreted to be self-controlling based on the pressure state. The release of hydrates and gas is important for its role in global climate and for its contribution to seafloor geomorphology. GHZ = gas hydrate zone. FGZ = free gas zone.

hydrate dissociation in slope failure [26, 27]; and (4) to estimate the potential role of catastrophic methane release on climate [28].

In this paper, we describe the results of the Fugro-McClelland Marine Geosciences Inc.'s piezoprobe and the Davis-Villinger Temperature/Pressure Probe (DVTP-P) pressure measurements made on ODP Leg 204 at ODP Site 1244, Hydrate Ridge, offshore Oregon, USA. The tools are designed to make rapid measurements of pressure and hydrologic properties in low permeability sediments. We analyze the results from both tools, compare their results, and comment on the *in situ* conditions by analyzing the data that most accurately represent the natural system.

PRESSURE MEASUREMENT

Direct pressure measurements are rare and expensive, but are required to advance research of submarine hydrodynamic systems. The ODP has historically relied on CORKs and ACORKs to monitor pressure, temperature, and fluid chemistry over many years. This characterizes the *in situ* conditions but the time and cost of acquiring data make the studies unrealistic for making robust and routine measurements beneath the seafloor.

An alternative approach to measuring *in situ* pressure is to use a penetration device. These measurements only take hours. Penetration devices that have been deployed in deep marine settings include free-fall penetration devices that sample pressure within a few meters of the seafloor. These include the Puppi [29, 30] and an early probe by Davis *et al.* [31]. A second class of instruments has been developed for use in boreholes. Two examples of these include the DVTP-P tool deployed on ODP Leg 190 [32] and the piezoprobe device [33–35].

The DVTP-P tool and the piezoprobe are similar devices. The tools have certain operational differences, with the key difference being the geometry of the tools (Figure 2). The tools induce a pressure pulse as they are inserted into sediments. The initial pressure response and its decay are defined by the insertion rate of the probe, the modulus of the sediment, and the bulk permeability of the sediment. The tool geometry coupled with the penetration rate dictate the spatial distribution of induced pressure; for a similar insertion rate, these tools produce different excess pore pressure distributions because of their different geometries. The pressure dissipation is used to infer *in situ* pressure and rock properties [34, 35].

The piezoprobe has a narrow probe that is 170 mm long including the short, tapered tip. The probe has diameter of 6.4 mm. A larger diameter shoulder assembly connects the probe to the drillstring [34] (Figure 2). At the tip of the probe, a porous element allows communication of pore fluid with the pressure transducer.

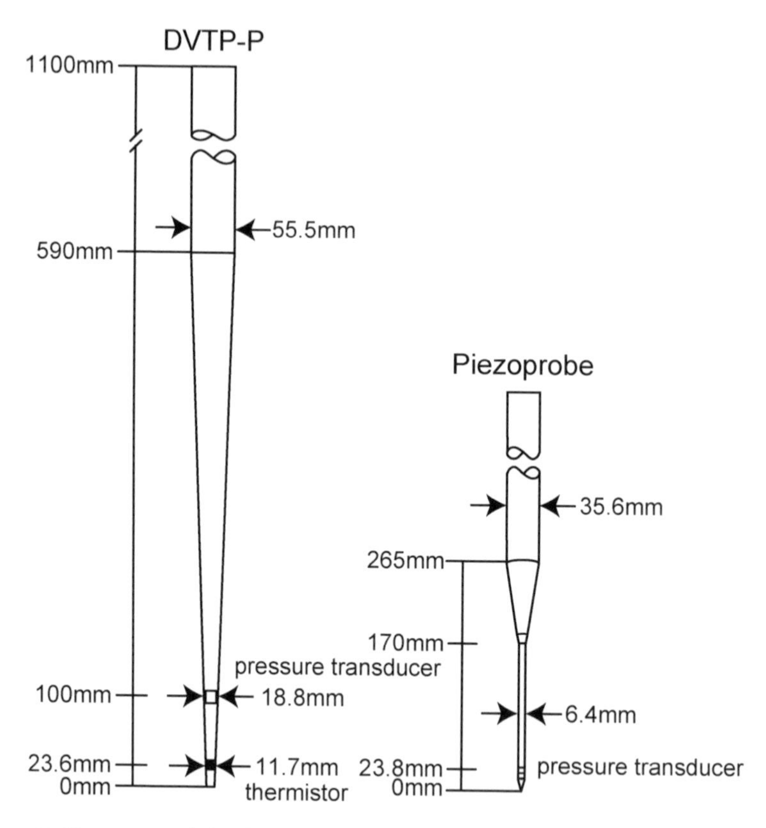

Figure 2. Geometry of the DVTP-P and the piezoprobe. Both tools have pressure transducers near their tip, but the tools have different geometries. The DVTP-P has a long, tapered cone that extends beyond the drillbit. DVTP-P geometry modified from [36]. The piezoprobe has a short, wide shoulder that is attached to a narrow lance where the pressure transducer is located. Geometry of piezoprobe based on [34]. The geometry of the probe and location of the pressure transducer affects the time required to accurately estimate *in situ* conditions.

The DVTP-P has a different geometry and thus a different pressure response. The DVTP-P has a longer and wider taper than the piezoprobe (Figure 2); its length is over twice that of the piezoprobe and the maximum diameter is almost twice that of the shoulder of the piezoprobe [36]. The pressure transducer is located farther from the probe tip than it is on the piezoprobe (Figure 2); this impacts the time required to interpret the *in situ* pressure and rock properties.

The tools have been designed to allow estimation of pressure and rock properties from the pressure data. The initial excess pressure during steady penetration can be related to the peak excess pressure and used to estimate the shear modulus of the sediments if conditions are undrained [37] or local permeability if partial drainage occurs [38]. After the tool insertion has ceased, the pressure dissipation allows estimation of the coefficient of consolidation [37, 39, 40], which can be used to infer permeability.

Penetration devices and long term monitoring stations will provide a full suite of pressure and rock property data beneath the seafloor that will increase our understanding of the sub-seafloor hydrologic system. The co-operative use of the devices will provide real-time and human-time scale data sets for understanding the dynamics of complex hydrodynamic systems. The data will also provide tests and calibrations of laboratory techniques used to interpret pressure, stress, and deformation. Many approaches have been used in the laboratory to estimate basin-scale pressures and rock properties from core samples [16, 41–43].

TEST SITE

ODP Site 1244 is located on Hydrate Ridge in 895.43 m of water (Figure 3). The presence of gas hydrate and free gas are interpreted based on a prominent bottom simulating reflector in seismic data [44]. One piezoprobe measurement was made at 53.66 meters below seafloor (mbsf) in Hole 1244C. This measurement was made in an interval of hemipelagic clay. A DVTP-P measurement was made in Hole 1244E at 52.6 mbsf in hemipelagic clay. Holes 1244C and E are located approximately 40 m apart. Site 1244 was dominated by hemipelagic clay with some turbiditic interlayers of silt and sand that find upward; the turbidite layers were most common and thickest between 69 and 245 mbsf. Below 245 mbsf, the lithology changes to indurated and fractured claystone with glauconite rich silt and sand interbeds.

At the depth of the piezoprobe and DVTP-P measurements, *in situ* porosity is between 61 and 64% (void ratio between 1.56 and 1.78), based on shipboard measurements of porosity from samples collected near the tool deployments (Figure 4). Porosity decreases downhole from 70% at the seafloor to just below 50% at 160 mbsf. The piezoprobe and DVTP-P deployments coincide to a depth where an increase in porosity is present (Figure 4).

Shipboard bulk density measurements were integrated to calculated the vertical hydrostatic effective stress (σ_{vh}') at Site 1244 (Figure 4); σ_{vh}' is the total overburden stress less hydrostatic fluid pressure ($\sigma_{vh}' = \sigma_v - u_h$). The vertical hydrostatic stress at the piezoprobe deployment depth is 0.331 MPa. At the DVTP-P deployment depth, σ_{vh}' is 0.327 MPa. Measurements on samples from ODP Site 892 (located near Site 1244 in Figure 3) establish the

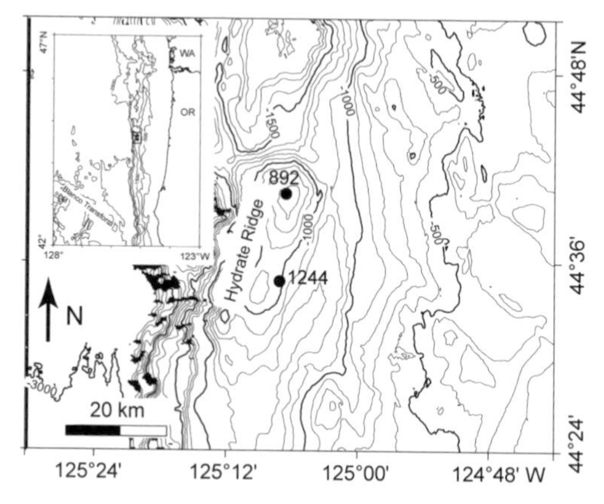

Figure 3. Hydrate Ridge is located offshore Oregon, USA (inset map). Bathymetry contour interval is 100 m. Site 1244 is located near the southern crest of Hydrate Ridge. Core samples from Site 892 on the northern crest of Hydrate Ridge were used to estimate *in situ* stress and pressures [16]. DVTP-P and piezoprobe measurements at Site 1244 will help to test these inferences based on consolidation behavior of the sediments from Site 892. Consolidation experiments from Site 1244 will also be completed to estimate pressure and stress for comparison to piezoprobe and DVTP-P measurements.

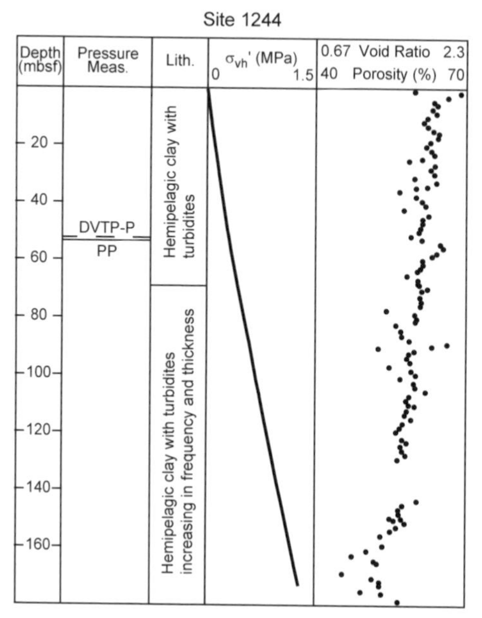

Figure 4. Summary of ODP Site 1244. PP = piezoprobe. Lithology is based on shipboard observations. Hydrostatic effective stress (σ_{vh}') is determined from density measured in Hole 1244C. Porosity from Hole 1244C is based on shipboard measurements and plotted on a linear scale; minimum and maximum void ratio are identified for reference.

permeability for the hemipelagic clay to be ~1×10^{-17} m^2 (range = 3.4×10^{-17}–8.5×10^{-18} m^2) at *in situ* stress [16].

PIEZOPROBE AND DVTP-P DEPLOYMENT

The piezoprobe was deployed on ODP Leg 204, Site 1244, Hole C on 14 July 2002. The deployment events for the test are described in Table 1 and the pressure history recorded by the piezoprobe is shown in Figure 5. We calculated the hydrostatic pressure (u_h) by assuming a fluid density of 1.024g/cm^3 (Table 2). We calculated the overburden stress (s_v) by integrating core porosity and density measurements (Figure 4; Table 2). Thirty minutes into the deployment (#2, Figure 5A) the tool reached the seafloor. Thereafter it was lowered 53 meters to the bottom of the borehole (#4, Figures 5A, 5B). The tool pressure when the probe is at or near the base of the hole is slightly greater than the estimated hydrostatic stress (Figure 5B). This could be due to poor tool calibration, a borehole fluid density greater than 1.024 g/cm^3 (due to sediment in the borehole or greater salinity), or additional borehole pressure resulting from pumping. The piezoprobe test lasted 45 minutes (Table 1, Figure 5A). An initial peak pressure of 10.29 MPa declined ultimately to 9.615 MPa (Table 3). This final pressure is 0.08 MPa greater than u_h (Figure 6).

The DVTP-P test lasted 33.5 minutes with a peak pressure of 10.55 MPa and a final pressure of 9.79 MPa (Table 3; Figure 6). Abrupt jumps in the

Table 1. Piezoprobe Deployment Log.

Event #	Time GMT	Time (minutes since deployment)	Event Description
1	7:32:34	0.565	Sitting in pipe—tip in water
2	8:09:22	37.365	Setting bit 7 meters from bottom
3	8:16:27	44.449	Lowering
4	8:22:11	50.182	Taking hydrostatic pressure
5	8:26:23	54.365	Pulled up 1.3 meters off of landing ring, now ~8 feet off bottom
6	8:27:13	55.215	Lowering bit down to 3.5 meters off bottom
7	8:36:35	64.582	Stopped pumping
8	8:38:22	66.365	Tagging bottom
9	8:39:31	67.515	Pushing
10	9:26:30	114.498	End of test—pulling
11	9:28:20	116.332	Coming to surface
12	9:45:42	133.699	At top of pipe

Table 2. Site Parameters.

	Site, hole	mbsf (meters)	Depth Below Sea Level (meters)	Overburden Stress, σ_v (MPa)	Hydrostatic Pressure, u_h (MPa)	Hydrostatic Effective Stress, σ_{vh}' (MPa)
Seafloor	1244C	0.0	895.43	8.995	8.995	0.0
Piezoprobe (7/14/02)	1244C	53.66	949.09	9.867	9.534	0.331
Seafloor	1244E	0	893.3	8.974	8.974	0
DVTP-P#2, Run 19 (8/19/02)	1244E	52.6	945.9	9.829	9.502	0.327

**calculations assume seawater density of 1.024 g/cm³.*

Table 3. Key Pressure Readings and Calculations.

Test	Duration of Dissipation (min)	Peak Pressure, u_i (MPa)	Pressure at End of Test (MPa)	Hydrostatic Pressure, u_h (MPa)	Inverse Time Prediction, u_{It} (MPa)	Final Pressure, u^* (MPa)
Piezoprobe	45	10.29	9.614	9.53	9.59	9.5
DVTP-P	33.5	10.55	9.79	9.598	9.71	–

Table 4. Nomenclature.

Symbol	Definition	Dimensions
k	hydraulic conductivity	L/T
R_2	radius at transducer	L
T	time factor	dimensionless
T_{50}	time factor at 50% dissipation	dimensionless
t	Time	T
t_{50}	time at 50% dissipation	T
u	pore pressure	M/LT^2
u_h	hydrostatic pressure	M/LT^2
u_i	peak pressure	M/LT^2
u_{It}	inverse time pressure estimate	M/LT^2
u^*	final pressure	M/LT^2
g_w	unit weight of water	M/L^2T^2
s'	mean effective stress	M/LT^2
s_v	overburden stress	M/LT^2
s_{vh}'	vertical hydrostatic effective stress	M/LT^2

pressure data approximately five minutes after insertion may have resulted from tension on the probe. The DVTP-P measurement was made in Hole 1244E, approximately 40 m from the piezoprobe test at Hole 1244C. It is reasonable to assume that these probes are sampling approximately the same material. Differences between the two measurements are: (1) the ini-

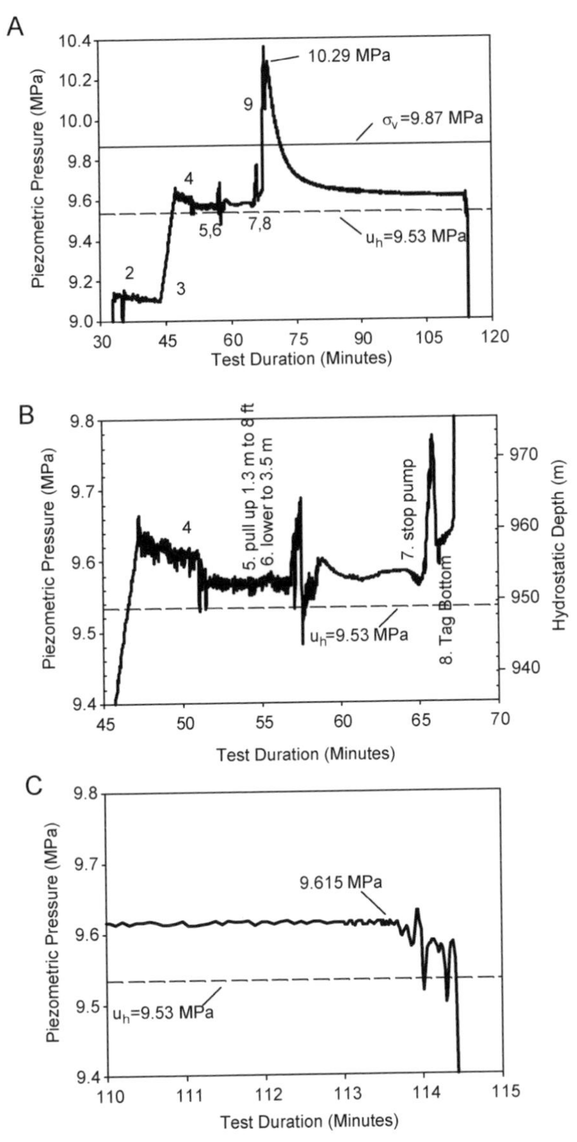

Figure 5. Pressure versus time for the piezoprobe deployment to 53.66 mbsf in Site 1244C on 14 July 2002. (A) Long term pressure record during the time the piezoprobe was near the seafloor. Hydrostatic pressure (u_h) and overburden stress (σ_v) for the depth of the piezoprobe penetration are shown. The piezoprobe deployment events are identified by number and are explained in Table 1. (B) Expanded view of the pressure prior to penetration. These data are generally used to estimate hydrostatic pressure. (C) Expanded view of the end of the dissipation profile. Piezoprobe pressure equilibrates to approximately the hydrostatic pressure.

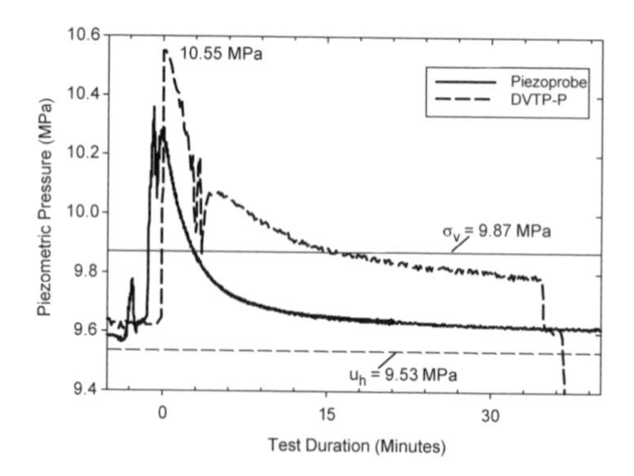

Figure 6. Comparison of piezoprobe pressure dissipation and DVTP-P pressure dissipation. The DVTP-P has a higher pressure than the piezoprobe during and after insertion; maximum pressure is 10.55 MPa for the DVTP-P versus 10.29 MPa for the piezoprobe. Hydrostatic pressure (u_h) and overburden stress (σ_v) at the piezoprobe deployment depth are shown for reference.

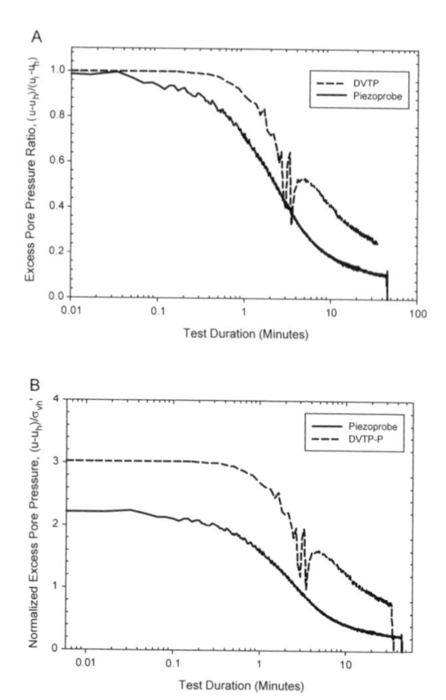

Figure 7. (A) Excess pore pressure ratio. Assuming that the *in situ* pressure is hydrostatic (u_h), the piezoprobe has dissipated 90% of its induced pressure while the DVTP-P has dissipated approximately 80% of its induced pressure. (B) Normalized pore pressure for the piezoprobe and DVTP-P. The piezoprobe has an initial pressure that is approximately two times the inferred *in situ* effective stress (σ_{vh}'). The DVTP-P pressure has a higher insertion pressure that only declines to approximately the *in situ* hydrostatic effective stress (normalized pressure = 1).

tial penetration pressure of the DVTP-P is significantly greater than the piezoprobe, (2) the DVTP-P pressure does not decline to as low a pressure as the piezoprobe pressure does, and (3) the DVTP-P pressure is dropping more rapidly than the piezoprobe pressure at the end of the test (Figure 6).

The excess pore pressure ratio (Figure 7A) is the pore pressure (u) normalized by the peak pore pressure (u_i) (Table 3). It is a useful way to measure the relative dissipation that has occurred. In this case, we have normalized the pressure relative to the hydrostatic pressure (u_h). From this plot is clear that the piezoprobe has dissipated significantly more relative to its peak pressure than the DVTP-P has dissipated. The normalized excess pore pressure (Figure 7B) is a measure of the magnitude of the pore pressure (u) relative to the hydrostatic effective stress (σ_{vh}'). The DVTP-P generates pore pressure three times greater than σ_{vh}', while the piezoprobe generates pressure only two times σ_{vh}' (Figure 7B).

PIEZOPROBE AND DVTP-P INTERPRETATION

We desire to interpret both the *in situ* pressure and hydraulic properties (e.g. permeability) from the piezoprobe test. Whittle *et al.* [34] propose that there is a characteristic dissipation curve associated with the piezoprobe and that given a soil model, permeability can be derived based on the following equation for normally consolidated clays,

$$T = \frac{\sigma' kt}{\gamma_w R_2^2} \quad . \tag{1}$$

T is the time factor and σ' is the mean effective stress. We have assumed $\sigma' = 0.67\sigma_{vh}'$. γ_w is the unit weight of water, k is the hydraulic conductivity, t is time, and R_2 is the radius of the piezoprobe at the shaft (35.6 mm). T_{50} is the time factor at 50% dissipation, while t_{50} is the absolute time at 50% dissipation. Whittle *et al.* [34] model T_{50} to be 1.72×10^{-3} for Boston Blue Clay with an overconsolidation ratio of 1.2.

To determine permeability we substitute T_{50} and t_{50} into Equation 1. However, to determine t_{50}, we must determine the final pressure, u^*. We assumed two values for u^*: 9.5 and 9.6 MPa. With these assumptions, two pore pressure ratio curves are generated and t_{50} is determined to be 120 and 165 seconds (gray and black solid curves, Figures 8A, 8B). The two u^* values yield hydraulic conductivities of 1.9×10^{-8} cm/sec and 1.4×10^{-8} cm/sec. These hydraulic conductivities equate to permeabilities of 1.96×10^{-17} m^2 and 1.43×10^{-17} m^2. The small variation in permeability suggests that the permeability is not very sensitive to the estimate of *in situ* pressure. These values are in the same range as those measured by [16].

To determine which of the proposed u^* values is appropriate, the curves are fitted to Whittle's normalized dissipation curve for Boston Blue Clay

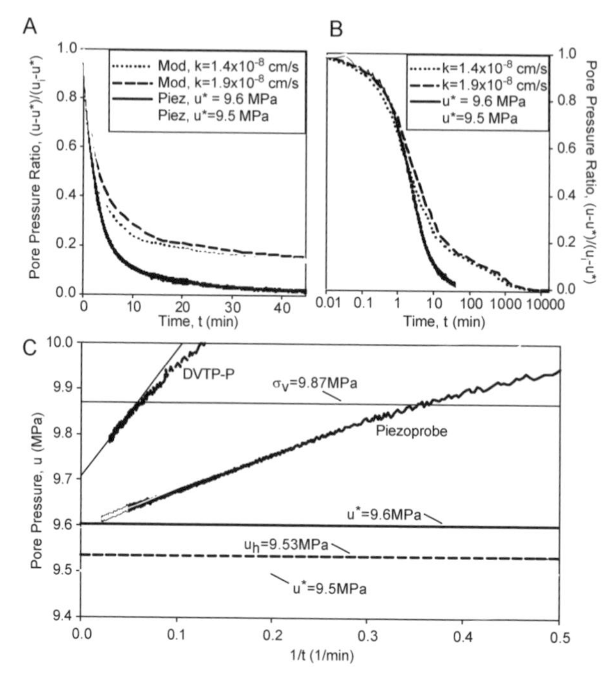

Figure 8. (A) Pore pressure ratio dissipation plots in linear time for the piezoprobe data assuming u^* equals 9.5 MPa (solid grey line) or 9.6 MPa (solid black line). Modeled pore pressure ratio dissipation plots in linear time are also shown assuming hydraulic conductivity is 1.4×10^{-8} cm/sec (dotted line) or 1.9×10^{-8} cm/sec (dashed line). Model results with either hydraulic conductivity are most similar to piezoprobe data assuming u^* equals 9.5 MPa (B) Pore pressure ratio dissipation plots in log time for the piezoprobe data assuming u^* equals 9.5 MPa (solid grey line) or 9.6 MPa (solid black line). Modeled pore pressure ratio dissipation plots in log time are also shown assuming hydraulic conductivity is 1.4×10^{-8} cm/sec (dotted line) or 1.9×10^{-8} cm/sec (dashed line). An *in situ* pressure of 9.5 MPa is consistent with model results. (C) Inverse time-pressure extrapolation to estimate *in situ* pore pressure for piezoprobe and DVTP-P data. DVTP-P data yield an estimate of 9.71 MPa whereas the piezoprobe data yield an estimate of 9.59 MPa. u^* values of 9.5 and 9.6 MPa used in (8A) and (8B) are shown for reference. u_h is also shown for reference.

(dotted and dashed lines, Figures 8A, 8B) [34]. In linear time (Figure 8A) and log time (Figure 8B), it is clear that with $u^* = 9.5$ MPa there is a much better fit of the modeled curve than with $u^* = 9.6$ MPa. $u^* = 9.5$ MPa is very close to the hydrostatic pressure ($u_h = 9.53$ MPa).

This prediction is compared to an inverse time extrapolation (Figure 8C). In this approach, measured pressures are plotted as a function of inverse time and the y-intercept is an estimate of the *in situ* pressure (u_{It}) [31, 33, 34]. We find u_{It} is 9.71 MPa for the DVTP-P and 9.59 MPa for the piezoprobe (Figure

8C). u_{lt} is an overestimate of the *in situ* pressure [45] and thus the u^* of 9.5 MPa derived from the piezoprobe is a reasonable estimate.

In summary, analysis of the piezoprobe data suggests that *in situ* pore pressure (9.5 MPa) is nearly hydrostatic (9.53 MPa) and *in situ* permeability is approximately 1.5×10^{-17} m^2. Over the time span of the piezoprobe test (45 minutes), 90% of the penetration-induced pore pressure was dissipated. It is important to recognize that the prediction of u^*, the *in situ* pressure, relies heavily on the model-based normalized pressure dissipation curve derived specifically for the piezoprobe geometry and specific soil parameters. Whittle *et al.* [34] describe in detail the fact that because of the geometry of the piezoprobe where a large diameter shaft overlies a narrow diameter probe, there is a shelf in the pressure data (Figure 8B, between 100 and 1000 min). The pressure induced by the large diameter shaft that reaches the pressure transducer causes this shelf.

A second primary result is that the DVTP-P pressures have not dissipated as much as the piezoprobe pressures either relative to their peak pressures (Figure 7) or in absolute pressure (Figure 6). This is not surprising because the radius of the DVTP-P is three times that of the piezoprobe at the pressure port and the DVTP-P continues to widen above the pressure port (Figure 2). The dissipation time is proportional to the square of the radius (Equation 1). Thus, based on a cylindrical probe geometry, t_{50} for the DVTP-P should be nine times that of the piezoprobe, but experimentally it is only five times as great (Figure 8B).

CONCLUSIONS

Deployment of the piezoprobe and the DVTP-P at Site 1244 of ODP Leg 1244 provided tests of both tools and estimates of *in situ* fluid pressures and rock properties. The piezoprobe data provide an estimate of *in situ* pressure equal to 9.5 MPa, which is nearly equal to the hydrostatic pressure at the depth of the experiment. The piezoprobe deployment and modeling of the results provides one of the first measurements of *in situ* permeability within the borehole. The dissipation data from the piezoprobe yield a permeability estimate of 1.5×10^{-17} m^2 for the hemipelagic clay; this is consistent with laboratory measurements (~1×10^{-17} m^2) on hemipelagic clay samples from nearby Site 892. The piezoprobe experiment only took approximately two hours from initial deployment until the tool was returned to the deck of the ship; 45 minutes of this time was the piezoprobe dissipation. The DVTP-P tool experiment, conducted in similar sediments, produced a significantly different pressure estimate. With 33 minutes of pressure dissipation, the DVTP-P pressure had dissipated to approximately the overburden stress and yields a pressure estimate of 9.71 MPa. Comparison of the DVTP-P and piezoprobe results suggest that a narrow diameter probe like the piezoprobe can be used to quickly and accurately determine *in situ* pressure and hydrologic properties of marine sediments.

The tests conducted on ODP Leg 204 are promising and suggest that future studies in marine geoscience and engineering can be strengthened with piezoprobe pressure and permeability observations. Continued use of the piezoprobe in sub-seafloor studies will expand research in a variety of geologic settings and will also provide real-time data that can be used to efficiently isolate regions of interest during emplacement of long term monitoring observatories.

ACKNOWLEDGEMENTS

This report would not have been possible without the contributions of Peter B. Flemings (Penn State Univ.), Frank R. Rack (Joint Oceanographic Institutions, Inc.), Gerhard Bohrmann (GEO-MAR), Anne M. Trehu (Oregon State Univ.), Derryl Schroeder (Ocean Drilling Program), Walter S. Borowski (Eastern Kentucky Univ.), Hitoshi Tomaru (Univ. of Tokyo), Marta E. Torres (Oregon State Univ.), George E. Claypool, Maarten Wouter Bart Vanneste (Univ. Tromso), Nathan L. Bangs (Univ. of Texas, Austin), Timothy S. Collett (USGS), Mark E. Delwiche (INEEL), Melanie Summit (Washington Univ.), Mahito Watanabe (Natl. Inst. of Adv. Industrial Sci & Tech.), Char-Shine Liu (Natl. Taiwan Univ.), Philip E. Long (Pacific Northwest Natl. Lab), Michael Riedel (Univ. of Victoria), Peter Schultheiss (GEOTEK Ltd.), Eulalia Gracia (CSIC), Joel E. Johnson (Oregon State Univ.), Alexei V. Milkov (Texas A&M), Barbara Teichert (GEOMAR), and Jill L. Weinberger (Scripps Inst. of Oceanography).

Assistance by J. Germaine (MIT) and A. Whittle (MIT) is deeply appreciated. Deployment of the piezoprobe was supported by the Ocean Drilling Program and the Department of Energy. This research used samples and/or data provided by t he Ocean Drilling Program (ODP). ODP is sponsored by the US National Science Foundation and participating countries under management of Joint Oceanographic institutions (JOI), Inc. Finally, the author recognizes the support of the Link Foundation through it's fellowship program.

REFERENCES

[1] D.W. Johnson, *The Origin of Submarine Canyons : A Critical Review of Hypotheses*, 126 pp., Hafner, New York (1939).

[2] P.A. Rona, "Middle Atlantic continental slope of United States; deposition and erosion", *AAPG Bulletin* 53(7), 1453–1465 (1969).

[3] K. Terzaghi, "Mechanism of landslides", in: *Application of geology to engineering practice: Berkey volume,* 83–123, Geological Society of America, New York (1950).
[4] E.G. Bombolakis, "Analysis of a Horizontal Catastrophic Landslide", in: *Mechanical Behavior of Crustal Rocks; The Handin Volume,* N.L. Carter, M. Friedman, J.M. Logan and D.W. Stearns, eds., 251–258, American Geophysical Union, Washington (1981).
[5] B. Dugan and P.B. Flemings, "Overpressure and Fluid Flow in the New Jersey Continental Slope: Implications for Slope Failure and Cold Seeps", *Science* 289, 288–291 (2000).
[6] B. Dugan and P.B. Flemings, "Fluid flow and stability of the US continental slope offshore New Jersey from the Pleistocene to the present", *Geofluids* 2(2), 137–146 (2002).
[7] A. Boehm and J.C. Moore, "Fluidized sandstone intrusions as in indicator of paleostress orientation, Santa Cruz, California", *Geofluids* 2(2), 147–161 (2002).
[8] W.C. Haneberg, "Groundwater Flow and the Stability of Heterogeneous Infinite Slopes Underlain by Impervious Substrata", in: *Clay and Shale Slope Instability,* W.C. Haneberg and S.A. Anderson, eds., 63–78, Geological Society of America, Boulder (1995).
[9] B.S. Hart, P.B. Flemings and A. Deshpande, "Porosity and pressure; role of compaction disequilibrium in the development of geopressures in a Gulf Coast Pleistocene basin", *Geology* 23(1), 45–48 (1995).
[10] N.L.B. Bangs, G.K. Westbrook, J.W. Ladd and P. Buhl, "Seismic Velocities from the Barbados Ridge Complex: Indicators of High Pore Fluid Pressures in an Accretionary Complex", *Journal of Geophysical Research* 95(B6), 8767–8782 (1990).
[11] D.E. Karig and G.F. Sharman, "Subduction and Accretion in Trenches", *Geological Society of America Bulletin* 86(3), 377–389 (1975).
[12] F.A. Dahlen, J. Suppe and D. Davis, "Mechanics of fold-and-thrust belts and accretionary wedges; cohesive Coulomb theory", *Journal of Geophysical Research* 89(12), 10087–10101 (1984).[13]
D. Davis, J. Suppe and F.A. Dahlen, "Mechanics of fold-and-thrust belts and accretionary wedges", *Journal of Geophysical Research* 88(2), 1153–1172 (1983).
[14] D.M. Saffer and B.A. Bekins, "Hydrologic Controls on the Morphology and Mechanics of Accretionary Wedges", *Geology* 30(3), 271–274 (2002).
[15] B.A. Bekins and S. Dreiss, "A Simplified Analysis of Parameters Controlling Dewatering in Accretionary Prisms", *Earth and Planetary Science Letter* 109, 275–287 (1992).
[16] K.M. Brown, "17. The Variation of the Hydraulic Conductivity Structure of an Overpressured Thrust Zone with Effective Stress", in: *Proceedings of the Ocean Drilling Program,* B. Carson, G.K. Westbrook, R.J. Musgrave and E. Suess, eds. 146, 281–289, Ocean Drilling Program, College Station (1995).
[17] E. Screaton, D. Saffer, P. Henry, S. Hunze and Leg 190 Shipboard Scientific Party, "Porosity loss within the underthrust sediments of the Nankai accretionary complex: Implications for overpressures", *Geology* 30(1), 19–22 (2001).
[18] E.E. Davis, K. Becker, T. Pettigrew, B. Caron and R. Macdonald, "CORK: A hydrologic seal and downhole observatory for deep-sea boreholes", in: *Proceedings of the Ocean Drilling Program, Initial Reports,* M.J. Mottl, E.E. Davis, A.T. Fisher and J.F. Slack, eds. 139, 42–53, Ocean Drilling Program, College Station (1992).

[19] E.E. Davis and K. Becker, "Formation temperatures and pressures in a sedimented rift hydrothermal system; 10 months of CORK observations, holes 857D and 858G", *Proceedings of the Ocean Drilling Program, Scientific Results* 139, 649–666 (1994).

[20] K. Becker, A.T. Fisher and E.E. Davis, "The Cork Experiment in Hole 949C: Long-Term Observations of Pressure and Temperature in the Barbados Accretionary Prism", *Proceedings of the Ocean Drilling Program Scientific Results* 156, 247–251 (1997).

[21] K.A. Kvenvolden, "Gas hydrates-geological perspective and global change", *Reviews of Geophysics* 31(173–187.) (1993).

[22] C. Ruppel, "Anomalously cold temperatures observed at the base of the gas hydrate stability zone on the U.S. Atlantic passive margin", *Geology* 25(8), 699–702 (1997).

[23] W.S. Holbrook, D. Lizarralde, I.A. Pecher, A.R. Gorman, K.L. Hackwith, M. Hornbach and D. Saffer, "Escape of methane gas through sediment waves in a large methane hydrate province", *Geology* 30(5), 467–470 (2002).

[24] R.D. Hyndman and E.E. Davis, "A mechanism for the formation of methane hydrate and Seafloor Bottom-Simulating reflectors by vertical fluid expulsion", *Journal of Geophysical Research* 97(B5), 7025–7041 (1992).

[25] X. Liu and P.B. Flemings, "Stress-limited gas column height in the gas hydrate system of Blake Ridge", in: *Proceedings of the 4th Annual International Conference on Gas Hydrates,* Yokohama, Japan, (2002).

[26] W.P. Dillon, W.W. Danforth, D.R. Hutchinson, R.M. Drury, M.H. Taylor and J.S. Booth, "Evidence for faulting related to dissociation of gas hydrate and release of methane off the Southeastern United States", in: *Gas Hydrates: Relevance to World Margin Stability and Climate Change,* J.-P. Henriet and J. Mienert, eds., Special Publications 137, 293–302, Geological Society, London (1998).

[27] W.P. Dillon, J.W. Nealon, M.H. Taylor, M.W. Lee, R.M. Drury and C.H. Anton, "Seafloor collapse and methane venting associated with gas hydrate on the Blake Ridge; causes and implications to seafloor stability and methane release", in: *Geophysical Monograph,* C.K. Paull and W.P. Dillon, eds. 124, 211–233 (2000).

[28] G.R. Dickens, M.M. Castillo and J.C.G. Walker, "A blast of gas in the latest Paleocene: Simulating first-order effects of massive dissociation of oceanic methane hydrate", *Geology* 25(3), 259–262 (1997).

[29] P.J. Schultheiss and S.D. McPhail, "Direct indication of pore-water advection from pore pressure measurements in Madeira abyssal plain sediments", *Nature* 320(6060), 348–350 (1986).

[30] R. Urgeles, M. Canals, J. Roberts and SNV "Las Palmas" Shipboard Party, "Fluid flow from pore pressure measurements off La Palma, Canary Islands", *Journal of Volcanology and Geothermal Research* 101, 253–271 (2000).

[31] E.E. Davis, G.C. Horel, R.D. MacDonald, H. Villinger, R.H. Bennett and H. Li, "Pore pressures and permeabilities measured in marine sediments with a tethered probe", *Journal of Geophysical Research* 96(4), 5975–5984 (1991).

[32] G.F. Moore, A. Taira, N.L. Bangs, S. Kuramoto, T.H. Shipley, C.M. Alex, S.S. Gulick, D.J. Hills, T. Ike, S. Ito, S.C. Leslie, A.J. McCutcheon, K. Mochizuki, S. Morita, Y. Nakamura, J.O. Park, B.L. Taylor, G. Toyama, H. Yagi and Z. Zhao,

Structural setting of the Leg 190 Muroto Transect, 14 pp., Texas A & M University, Ocean Drilling Program, College Station, TX (2001).

[33] R.M. Ostermeier, J.H. Pelletier, C.D. Winker and J.W. Nicholson, "Trends in Shallow Sediment Pore Pressures—Deepwater Gulf of Mexico", *Society of Petroleum Engineers* SPE/ IADC 67772 (2001).

[34] A.J. Whittle, T. Sutabutr, J.T. Germaine and A. Varney, "Prediction and interpretation of pore pressure dissipation for a tapered piezoprobe", *Geotechnique* 51(7), 601–617 (2001).

[35] A.J. Whittle, T. Sutabutr, J.T. Germaine and A. Varney, "Prediction and interpretation of pore pressure dissipation for a tapered piezoprobe", in: *Offshore Technology Conference,* Houston, (2001).

[36] D. Schroeder, personal communication (2002).

[37] M.F. Randolph and C.P. Wroth, "An Analytical Solution for the Consolidation Around a Driven Pile", *International Journal for Numerical and Analytical Methods in Geomechanics* 3, 217–229 (1979).

[38] D. Elsworth, "Dislocation Analysis of Penetration in Saturated Porous Media", *Journal of Geotechnical Engineering* 117(2), 391–408 (1991).

[39] R.C. Gupta and J.L. Davidson, "Piezoprobe Determined Coefficient of Consolidation", *Soil and Foundations* 26(3), 12–22 (1986).

[40] D. Elsworth, "Analysis of Piezocone Dissipation Data Using Dislocation Methods", *Journal of Geotechnical Engineering* 119(10), 1601–1623 (1993).

[41] D.E. Karig and G. Hou, "High-stress consolidation experiments and their geologic implications", *Journal of Geophysical Research* 97(1), 289–300 (1992).

[42] D.M. Saffer, E.A. Silver, A.T. Fisher, H. Tobin and K. Moran, "Inferred Pore Pressures at the Costa Rica Subduction Zone: Implications for Dewatering Processes", *Earth and Planetary Science Letters* 177, 193–207 (2000).

[43] W.J. Winters, "Stress History and Geotechnical Properties of sediment from the Cape Fear Diapir, Blake Ridge Diapir, and Blake Ridge", in: *Proceedings of the Ocean Drilling Program, Scientific Results,* C.K. Paull, R. Matsumoto, P.J. Wallace and W.P. Dillon, eds. 164, 421–429, Texas A & M University, Ocean Drilling Program, College Station, TX (2000).

[44] G. Bohrmann, A.M. Trehu, J. Baldauf and C. Richter, "Drilling Gas Hydrates on Hydrate Ridge, Cascadia Continental Margin: Leg 204 Scientific Prospectus", http://www- odp.tamu.edu/publications/, (2002).

[45] A.J. Whittle, personal communication (2002).

Air-Sea Momentum Measurement System for the Surfzone

Harry C. Friebel

Davidson Laboratory

Stevens Institute of Technology

Castle Point on Hudson

Hoboken, NJ 07030

Research Advisor: Dr. Thomas. O. Herrington

ABSTRACT

A high frequency air-sea momentum system has been developed consisting of the following components: 4 high-frequency salt-water wave wires, a high-frequency 3–D sonic anemometer, an accelerometer, a mass storage datalogger, a rechargeable battery pack and an onshore computer. The system layout is designed such that the wave transformation through the surf zone can be measured at different locations through the saltwater wave wires while also collecting wind fluctuation measurements over one particular wave wire. The system was evaluated during an experiment conducted in June of 2002 at Sandy Hook, NJ to measure turbulent fluxes within the surf zone on both sides of the air-sea interface. A total of 22, 30–minute data sets were collected over a three-day period at a sampling frequency of 10 Hz.

INTRODUCTION

Van de Voorde *et al.* [1] have observed that over a small time period, strong offshore winds can quickly dissipate energy from surface gravity waves. Many physical processes are responsible for wave energy decay in the surf zone—wave shoaling, breaking, refraction, diffraction, bottom friction, currents, topography and wind all play an important role in wave transformation. Of particular interest, the air-sea momentum flux is an important parameter when investigating wave transformation, shoaling and breaking within the surf zone. Direct measurement of these fluxes involves sampling the turbulent fluctuations of the wind velocity over a wide range of scales, and correlating them with concurrent and co-located measurements of the turbulent fluctuations of the vertical velocity [2]. The momentum flux vector, , is given by

$$\tau = -\rho(\overline{u'w'}) = \rho u_*^{\,2}, \quad (1)$$

where ρ is the air density, u' and w' are the turbulent components of the horizontal and vertical velocities, respectively, the overbar refers to time averaging over a suitable period (order 30 min) and is the friction velocity.

Direct measurements of the turbulent fluxes on both sides of the air-sea interface are quite difficult to obtain and hence bulk methods have been developed to estimate the fluxes from easier obtainable data. However, Volkov [3] recognized that applying the Monin-Obukhov method of turbulent scaling was inappropriate when swell waves exist. Swell waves may also prevent the use of the roughness parameter and wave age as noted by Donelan *et al.* [4]. Donelan suggest this may be due to the presence of a wave boundary layer (WBL) during swell conditions and as a result Monin-Obukhov theory does not apply in the WBL [2]. Shoaling and breaking waves have a constantly evolving wave face and consequently constantly changing surface roughness, so again the application of the Monin-Obukhov theory may not be appropriate for breaking waves. It is presently unclear how the momentum transfer between the air and sea is altered when significant angles exist between the wind and sea, as often occurs in the nearshore where strongly refracted low frequency swell is commonly at an angle to the wind.

To gain a better understanding of how offshore (opposing) winds can extract energy from the incident wave field, a unique high frequency air-sea momentum system has been developed. The system was deployed in the surf zone for three days in June 2002 as part of an experiment to quantify air-sea momentum transfer when the wind and wave direction were at oblique angles. The system obtained measurements in the nearshore via a high-resolution (60 Hz) 3–D sonic anemometer and high-frequency saltwater wave wires developed with the support of the Link Foundation. The designed wave wires measure the sea surface fluctuations in the short-gravity to gravity-capillary wave bands at a frequency of 10 Hz. The high-frequency wave gauging system is integrated with the sonic anemometer through a unique data acquisition system consisting of wave wire conditioners and a high-frequency datalogger. The advantage of this system is that direct measurements of the turbulent fluctuations on both sides of the air-sea interface are obtained and consequently the application of bulk methods is not necessary, allowing for the investigation of the effect of local wind stress on shallow water breaking waves.

HIGH FREQUENCY AIR-SEA MOMENTUM SYSTEM

In order to quantify the air-sea momentum exchange at highly-oblique wind and wave angles, a system capable of measuring high-frequency turbulent fluctuations on both sides of the air-sea interface has been developed. The system is comprised of the following components: 4 high-frequency salt-water wave wires, a high-frequency Campbell Scientific 3–D sonic anemometer, an Analog Device accelerometer, a Campbell Scientific CR23X datalogger, a rechargeable battery pack and an onshore laptop computer. The system is designed such that the wave transformation through the surf zone can be measured at different locations through the use of newly developed saltwater wave wires while also collecting wind fluctuation measurements over one particular wave wire. The system allows for data retrieval from a remote access computer so that the quality of data can continually be evaluated and no more than one half hour of collected data would be lost if the system goes down.

Cross-shore wave measurements are obtained by using resistance wave wires deployed in the nearshore in mean water depths of approximately 5 to 15 feet. The wave wires are constructed using corrosive resistant materials, and are waterproofed for long-term reliability. The wetted components are constructed of 304 stainless steel and PVC plastic. The active sensing length of each wire is approximately 68 in. The probes are designed to install vertically with two conduit hangers attaching the upper and lower insulating PVC blocks to support poles (Figure 1). The conduit hangers secure the wave wire to a 1½ in. steel support pole that extends above the water surface and is jetted approximately 6–8 ft. into the seabed. All internal wiring is waterproofed using a silicone based electrical potting compound. The sensing element is a .0625 in. diameter 316 stainless steel wire, stretched between the upper and lower PVC insulating blocks and is made to be replaceable in the field, in the event of damage. Each wire uses two .375 in. diameter stainless steel rods to provide a ground connection into

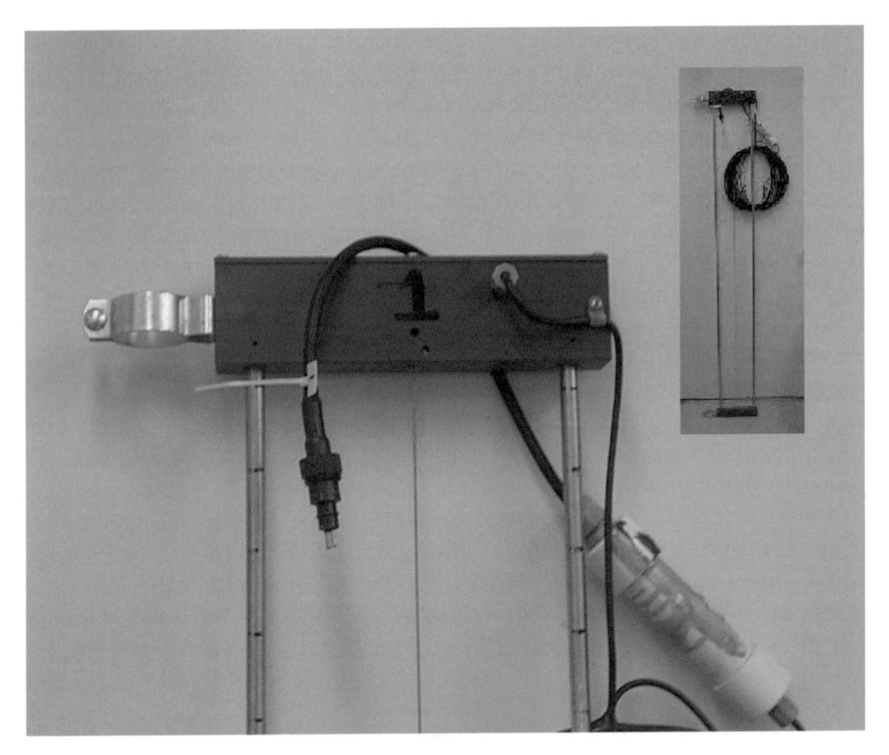

Figure 1. High Frequency Salt Water Wave Wire.

the water surface. The use of two ground rods, with the sensing element between them, provides a balanced current field through the water. Each wave wire is hardwired to RG-58C/U coaxial cable with a noncontaminating PVC jacket to prevent water from seeping into the dielectric of the cable. (Weiss,[5]) The end of the cable is connected to a male SeaCon waterproof connector through an in-house constructed PVC waterproof connection. The male SeaCon connector attaches to a female SeaCon connector that is hard wired to the datalogger located within a waterproof pelican box.

The waterproof Pelican 1600 case contains both the signal conditioners for the wave wires and the Campbell Scientific CR23X datalogger (Figure 2). The signal conditioner unit, housed underneath the display panel of the pelican box, contains five separate channels of signal conditioning. All five channels share a common ground, but retain separate power supplies and excitation sources. In this way, a power supply or other component failure will affect only one channel of the five, permitting continued operation in the field.

The wave wires, sonic anemometer, and accelerometer are hard wired to a datalogger that is mounted within the pelican case display panel. The datalogger is an 11 to 16 Volt-powered programmable measurement and control system designed to measure 12 differential or 24 single-ended analog data channels. The programmable module provides sensor measurement, communication, data reduction, data/program storage and control functions. A battery-backed clock assures accurate timekeeping. The system is designed to perform multitasking operations, allowing for simultaneous communication and measurement functions. Systems tasks are initiated in sync with real-time sampling frequencies of up to 100 Hz. Measurement rates up to 1.5 kHz are possible over short intervals using burst measurements. Data and programs are stored either in non-volatile flash memory or battery backed SRAM. The 1 M memory stores 500K data values. (Herrington, [6]) The CR23X is powered by two rechargeable 12 VDC batteries that can power the system for approximately five days.

For air-sea momentum flux measurements, the CR23X datalogger is programmed to record the Julian day time stamp, four voltage outputs from the wave wires, the five channels (u, v, w, speed of sound and diagnostic) of the anemometer and the three channels (a_x, a_y a_z) of the accelerometer, simultaneously at a frequency of 10 Hz. Datalogger programming is achieved by the windows based program (PC208W) developed by Campbell

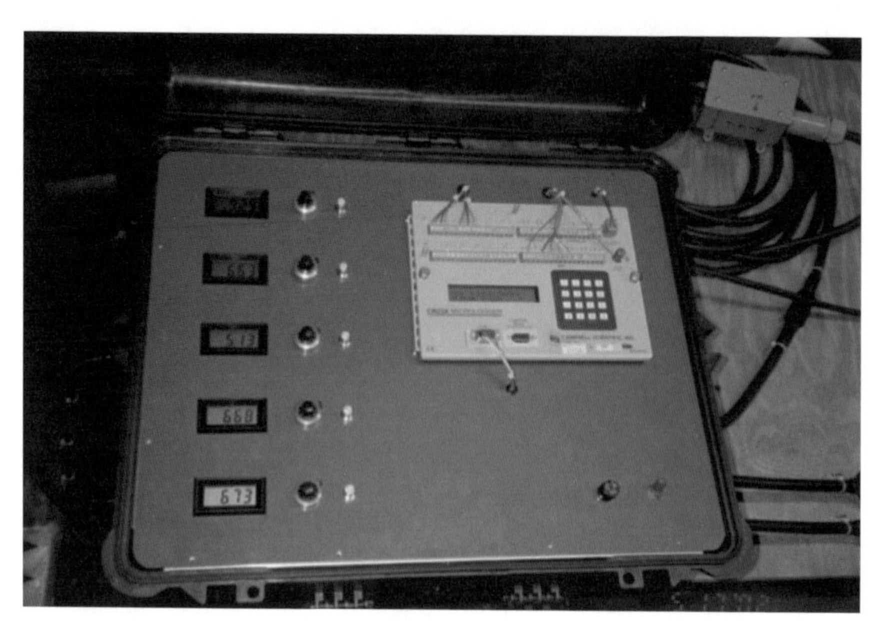

Figure 2. Pelican Box with Signal Conditioners.

Scientific, Inc. All programming, data retrieval and storage, processing and data transfer is controlled by a Pentium computer. Due to limitations on the memory capacity of the datalogger, to be discussed further, the data collection time is limited to 30 minutes. The recorded data is downloaded immediately after the end of the sampling interval, taking approximately 20 minutes to transmit the data from the CR23X datalogger onshore through a double armored steel data/power cable, leaving just 10 minutes to verify all instrumentation are working correctly before the collection cycle repeats. A portable zip drive is attached to the laptop and backs up the downloaded data after transmittal. The laptop and portable zip drive are powered from a portable Honda generator.

The high-frequency 3-D sonic anemometer and accelerometer are co-located on a single instrumentation mast (Figure 3). In order to produce

Figure 3. Central Pole Deployed.

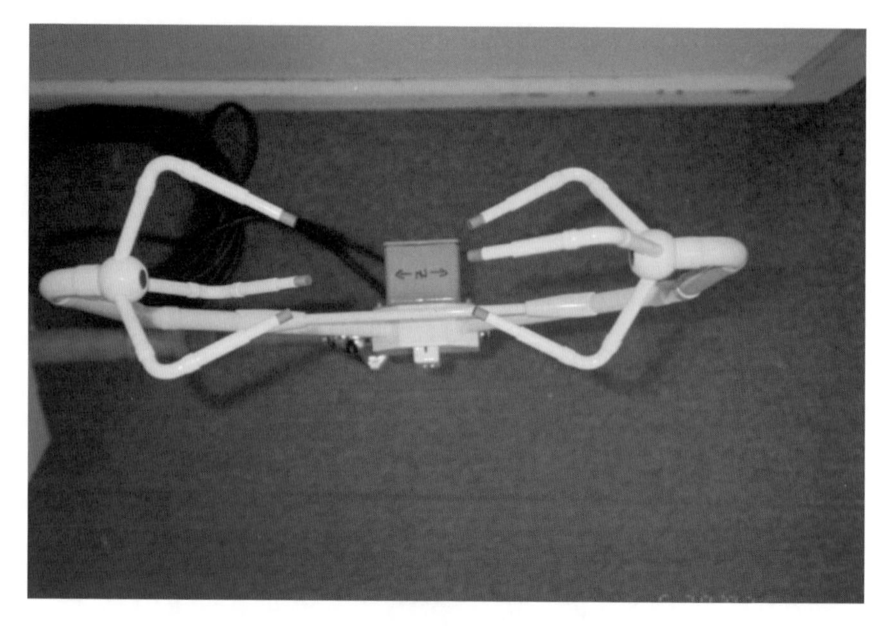

Figure 4. Anemometer with Accelerometer Attached.

exact synchronization with the other instruments, the 60 Hz sonic anemometer is programmed to measure the three orthogonal wind component velocity (u, v, w) fluctuations at 10 Hz. The anemometer has a 10 cm vertical measurement path and uses acoustic pulses from 3 acoustic prongs to measure wind components while correcting for cross wind effects. The accelerometer is housed in a PVC waterproof enclosure and positioned on the horizontal axis of the sonic anemometer to measure the motion of the instrument mast (Figure 4). The central pole is expected to oscillate, due to wind and wave force impacts. This oscillation will in turn affect the sonic anemometer measurements and will need to be extracted from the data. The accelerometer sensor will transform the wind velocity measurements to a stationary reference frame. Both the anemometer and accelerometer are hardwired to the datalogger through waterproof SeaCon connectors attached to the outside of the pelican box and are powered from the CR23X datalogger.

TEST AND OPERATION

Calibration and Noise

Calibration and sensitivity test of the wave wire system were conducted during May 2002 in a saltwater tank, located in the Davidson Laboratory.

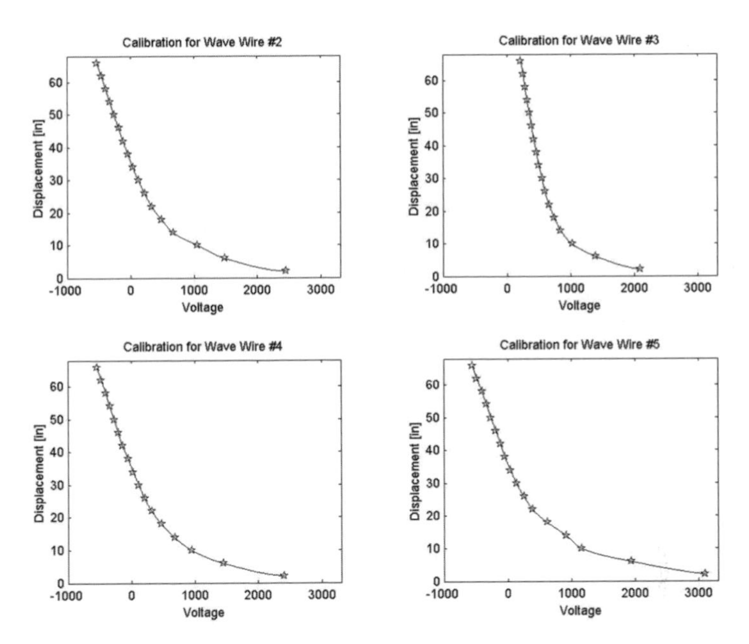

Figure 5. Wave Wire Calibration Data.

During calibration, the entire deployed air-sea momentum system is mimicked in the laboratory, by collecting data through all instruments during the wave wire calibration process. This is done to insure the datalogger records at the proper frequency, the datalogger outputs the proper voltage to each wire while contending with the voltage draw from the other instruments and to verify the datalogger does not skip scans. During calibration, each wire is lowered into the tank in increments of two inches, starting from the bottom of each wire and the voltage is recorded. Once the wire is completely submerged, the wire is returned to the midpoint and initial calibration point to confirm the previously recorded voltage readings. This process is done for all wires, under three different salinities of 25, 30 and 35 ppt. The voltages are then plotted and a piecewise cubic Hermite polynomial interpolates the voltages to a submerged distance of the wave wire. The interpolating polynomials are plotted below (Figure 5).

In order to assess the level of background signal noise in the system, all four wave wires are partially submerged to a noted particular depth and the voltage outputs are recorded. The system collects data for 30 minutes and the final voltage recordings are compared to the initial, investigating for voltage drift. A small noise signal was found in all channels sampled.

SAMPLING FREQUENCIES

Before deployment, the maximum sampling frequency of each instrument needed to be determined. The high-frequency air-sea measurement system is comprised of two separate systems collecting data through the datalogger at the top of every hour; the wave wires and the sonic anemometer,. The measurement program collects seventeen channels of data; a program stamp, the Julian day, the hour and minute, the second with tenths precision, wave wire #1, wave wire #2, wave wire #3, wave wire #4, wave wire #5, accelerometer x-direction, accelerometer y-direction, accelerometer z-direction, anemometer x-direction, anemometer y-direction, anemometer z-direction, the speed of sound and an anemometer diagnostic counter. By testing, a sampling frequency of 10 Hz was found to be within the capabilities of the datalogger. At this frequency, no scans were missed and it is considered adequate to measure the sea surface fluctuations in the short-gravity to gravity-capillary wave bands. Measurement frequencies higher than 10 Hz caused the datalogger to skip scans.

Datalogger memory capacity also limits the sampling interval. At a sampling frequency of 10 Hz, there are 58 bytes of data recorded per scan, or 34800 bytes per minute. The CR23X datalogger has 1664K of available memory, limiting the system to 48 minutes of data collection time before the memory fills and starts to overwrite. Applying a factor of safety and identifying that it takes approximately 20 minutes to download the data from previous testing, a 30–minute data collection window starting at the top of every hour has been established. For all practical purposes, a PC running PC208W is required to retrieve the data and act as the mass storage device.

Coastal Field Test

A field test of the high frequency air-sea momentum system was conducted in Monmouth County, NJ at the Gateway National Recreational Area, located on Sandy Hook. Sandy Hook was chosen as the experiment location for its ideal geological formation; it is a unique, low lying sand peninsula with open water exposure to the east and west that allows for maximum wind exposure, with minimum disturbance from the offshore direction (Figure 6).

In May 2002, a preliminary beach survey was conducted to determine the deployment locations for the wave wires and sonic anemometer. In June 2002, the air-sea momentum system was deployed in the nearshore from the Davidson Laboratory's R/V Phoenix. At each deployment location, a 1½ in. steel pole was jetted into the seafloor via a water pump located onboard the research vessel. Once the pole was properly positioned

Figure 6. Gateway National Recreational Area, Sandy Hook, NJ.

in the seabed, a diver and a deck hand attached a wave wire to the pole. Two wires were positioned well seaward of the breaker line, a third wire (central) was positioned at the breaker line, and the final wire positioned landward of the breaker line. The central pole supported the sonic anemometer, the accelerometer and the waterproof Pelican 1600 case containing the datalogger and signal conditioners. The central pole was positioned such that the sonic anemometer and datalogger were not directly subjected to constant sea spray.

Once all four poles and wave wires were installed, a steel messenger cable is strung along the seabed between the poles. The wave wire data cables were then zip tied down the steel pole and messenger cable up to

Figure 7. ADV with Rechargeable Battery Pack.

the central pole. The cables were then attached to the pelican box through waterproof SeaCon connectors. The messenger cable serves to reduce stress along the data cables.

A Sontek Acoustic Doppler Velocimeter (ADV) ocean wave gage was positioned directly under the central wave wire (Figure 7). The ADV was purposely positioned in this location to verify and correlate the central wave wire measurements and to provide predominant inshore wave directions. The ADV recorded at 10 Hz for the first twenty minutes every hour, measuring both water pressure and the 3 orthogonal components of velocity. The ADV contains an internal battery power source and was mounted on a mooring built by the Davidson Laboratory that also contained the rechargeable battery pack for powering the high-frequency air-sea momentum system.

To complete the deployment a double armored steel data/power cable was hand carried out through the surf zone and attached to the pelican box through a waterproof SeaCon connector. The shore end of the cable was attached to a communications port on a Pentium PC computer. All programming, data retrieval and storage, processing and data transfer was controlled by the shore-based PC via the double armored steel data/power cable. The cable was strain relieved at both ends to alleviate any stresses on the cable. Once the system was confirmed to be powered and working, the

Figure 8. Deployed Air-Sea Momentum System.

data measurement program was initiated, sampling the anemometer, accelerometer, and 4 wave wires at 10 Hz, for 30 minutes each hour (Figure 8).

DATA

A total of 22, 30–minute data sets were collected over a three-day period in June of 2002. Wave wires four (WW4) and five (WW5) were not connected to the air-sea momentum system until day two, due to time constraints during installation. This results in 16 data sets where all four deployed wave wires were connected to the system. These 16 data sets were inspected for completeness by examining for skipped scans, active wave wire length relative to the water level and for outliers. One data set is considered incomplete due to WW1 being submerged below the water level. Six other data sets are incomplete due to WW4 and WW5 skipped scans. The author believes that the waterproof SeaCon connectors may have become completely submerged and moisture may have intruded and hence interrupted the output voltage signal. Nine data sets out of the 16 data sets

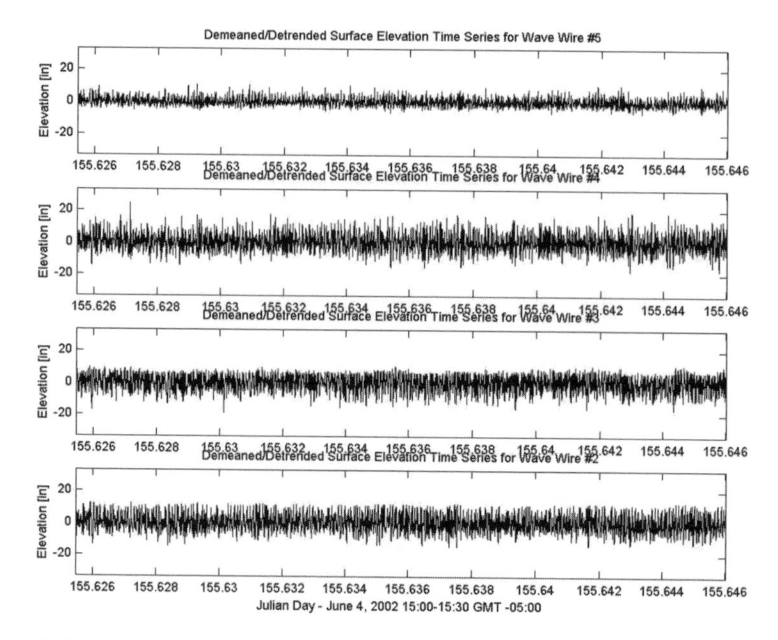

Figure 9. Wave Wire Data Plots.

fit the criteria for a complete data set. The wave wire raw data sets were first demeaned and then detrended to remove the tides from the signals.

To date, only the raw time series have been processed. However, the recorded data clearly shows the capabilities of the air-sea momentum system. The data for WW5 through WW2 during run number two are plotted below in Figure 9. WW5 is located the farthest offshore and the wave wire numbers count down moving inshore.

One interesting visual feature in the blow up data plot (Figure 10) is the wave trough that passes WW4 just after Julian Day 155.6259. The trough passes WW4 at approximately 155.62591, passes WW3 three seconds later at 155.62595 and WW2 six seconds later at 155.62598. Figure 10 visibly shows the quality and resolution of the processed data from the air-sea momentum system.

The anemometer and accelerometer raw data sets were also demeaned and are shown in Figure 11 and Figure 12 respectively. The accelerometer data plots shows acceleration spikes and noise that needs to be extracted to properly determine the accelerations of the anemometer. The presence of noise was confirmed in the previously noise test section.

An interesting feature in a blow up of the anemometer data plot (Figure 13) is the rapid wind decrease (direction change) after Julian Day 155.638 in the x and y -directions (directions relative to the anemometer orienta-

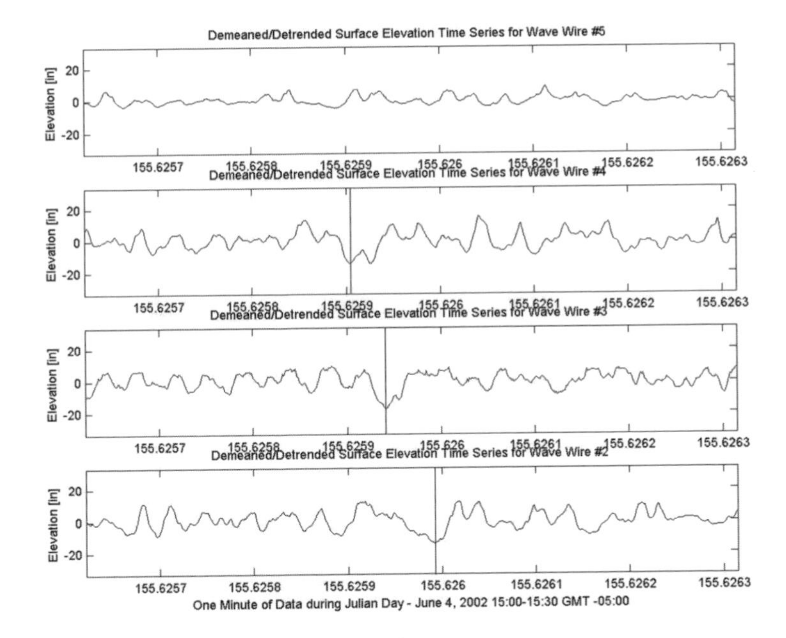

Figure 10. Wave Propagation Across Surf Zone.

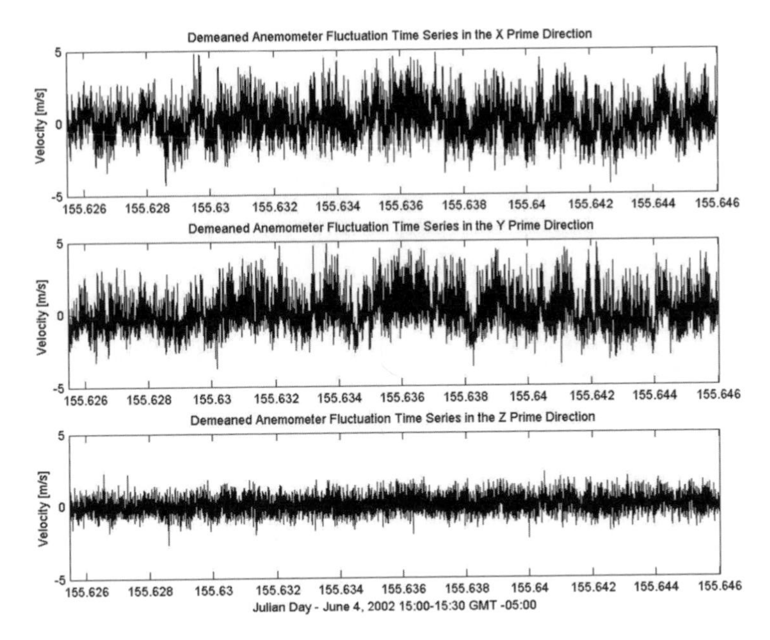

Figure 11. Anemometer Data Plots.

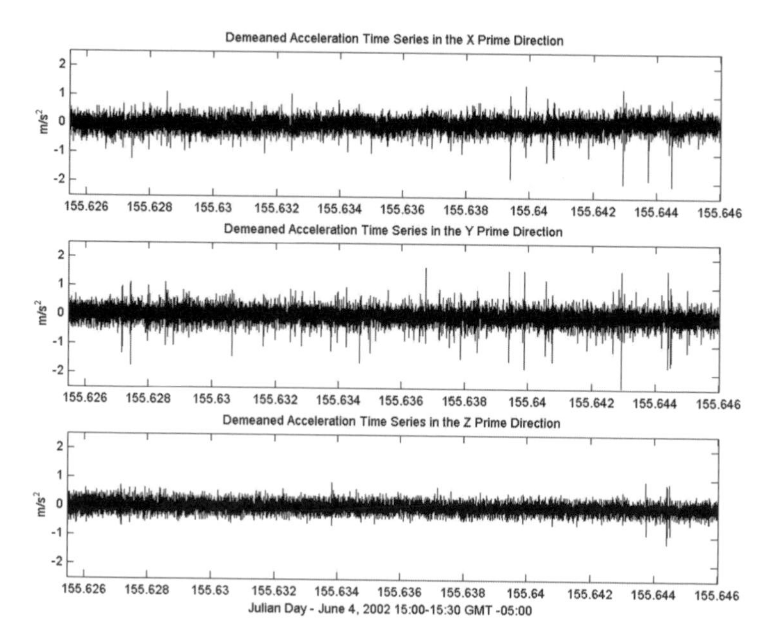

Figure 12. Accelerometer Data Plots.

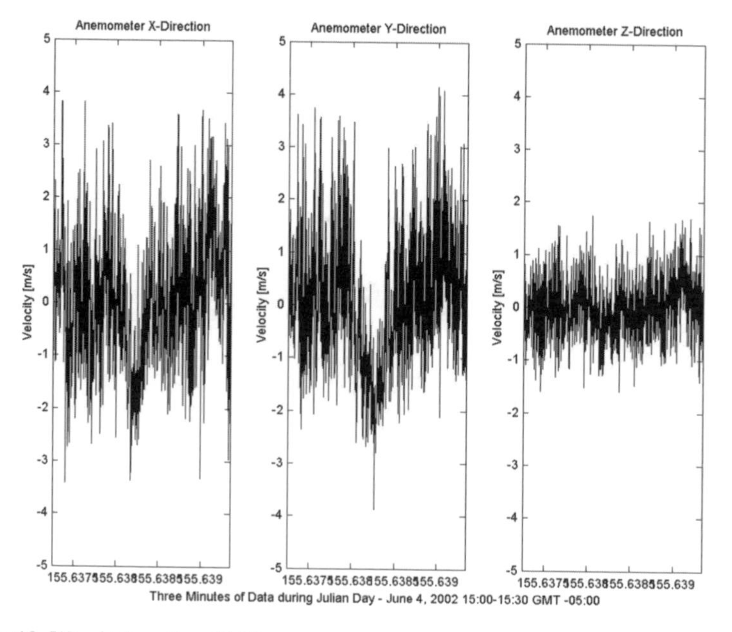

Figure 13. Wind Direction Shift.

tion). The average z -direction fluctuations are three to four times less in magnitude than the x and y -direction fluctuations as to be expected.

CONCLUSION

The prototype air-sea momentum system has proven itself to be a reliable tool in collecting momentum fluxes on both sides of the air-sea interface under harsh conditions. Data sets of high quality and frequency have been collected in the surf zone over an extended period of time. The data will be used to develop a better understanding of the air-sea momentum exchange in the nearshore region under oblique wind and wave angles. The analysis of the collected data should help improve our ability to model and predict the nearshore wave climate.

ACKNOWLEDGMENTS

This research would not be possible without the generous support of the Link Foundation. The author acknowledges the constructive guidance and encouragement of his advisor and friend Dr. Thomas O. Herrington. I also want to thank Dr. Michael Bruno, director of the Davidson Laboratory, for his continued support and financial assistance. With gratitude, I thank the following members of the Davidson Laboratory: Dr. Alexander Benilov, Brian Fullerton, Bob Weiss, Doug Medding, Pat Burke, Alex Chia, Ben Shaw, and Jeremy Turner. I would finally like to thank Mr. Robert Miskewitz, the National Park Service (Gateway National Recreational Area Unit), and the New Jersey Marine Sciences Consortium.

REFERENCES

[1] N.E. Van de Voorde and S.P. Dinnel, "Observed directional waves spectra during frontal passage" *Journal of Coastal Research* 14 (1), 337–346 (1998).

[2] W.M. Drennan, H.C. Graber, D. Hauser and C. Quentin, "On the wave age dependence of wind stress over pure wind seas" *Journal Geophysical Research* (accepted 2001).

[3] Y.A. Volkov, "Turbulent flux of momentum and heat in the atmospheric surface layer over a disturbed surface," *Izv., Atmospheric. and Oceanic Physics Series* 6, 770–774 (1970).

[4] M.A. Donelan, F.W. Dobson, S.D. Smith, and R.J. Anderson, "On the dependence of sea surface roughness on wave development" *Journal Physical Oceanography* 23, 2143–2149 (1993).

[5] R.E. Weiss, "Operating instructions and technical descriptions of the SSP-01–3D swash and sand probe" *Technical Report SIT-DL-95–9–2726*, (1995).
[6] T.O. Herrington, M.S. Bruno and K.L. Rankin, "The New Jersey coastal monitoring network: a real-time coastal observation system, *Journal Marine Environmental Engineering* 6 (1), 69–82, (2001).

Experimental Flow Characterization of Anguilliform Swimming Motion

Anna Pauline Miranda Michel

Department of Ocean Engineering

Massachusetts Institute of Technology

77 Massachusetts Avenue

Cambridge, MA 02139

Research Advisor: Professor Michael S. Triantafyllou

ABSTRACT

Data is presented for experimental studies of an anguilliform swimming body. Experimental measurements were performed to examine the flow characteristics about the swimming body. Past studies on undulatory swimming motion have shown that fish exhibit vorticity control as the undulatory motion of the fish combined with the flapping tail motion creates a jet wake. In addition, previous work has shown that unsteady motions are effective for controlling flow. Two-dimensional waving plate studies have shown turbulence reduction when the phase speed (Cp) increased compared to the freestream speed (Uo).

A 3D flexible snake-like robot was designed and tested in the MIT Marine Hydrodynamics Water Tunnel. Laser Doppler Velocimetry (LDV) was used to study the near body turbulence levels. Near-body turbulence studies at two locations along the length of the snake showed turbulence reduction for Cp/Uo = 1.2 when the snake was in the trough position. In the crest position, the snake showed a turbulence maximum at Cp/Uo = 1.2. The trough data supports previous 2D studies of waving plates. The crest data however shows opposite results.

INTRODUCTION

This work examines the hydrodynamic characteristics of a biomimetically designed swimming sea snake in a recirculating water tunnel. A snake robot was developed to mimic anguilliform swimming motions. This work centers about two main goals: 1) to elucidate the underlying hydrodynamic mechanisms of flow about swimming bodies similar to sea-snakes, and 2) to gain a better understanding of the anguilliform mode of propulsion and how it can be used in biomimetic underwater vehicle design.

Biomimetic Design

The field of biomimetics applies ideas from nature to solve engineering problems with the belief that plants and animals have evolved optimal solutions for survival. In this study, biomimetics is used in the area of robotics to study the undulatory motion of a swimming sea snake. Specific characteristics of the sea snake's body shape and swimming motion have been replicated in the model snake. Natural selection has enabled fish and other aquatic swimmers to evolve to be highly efficient, though not necessarily optimal swimmers. Inspiration from them can be applied to the field of robotics to improve underwater vehicle design. One example of an application is in the development of AUVs, autonomous underwater vehicles. As the use of AUVs increases, longer mission duration and efficiency will be required. Applying an efficient swimming motion could produce improved propulsion over that of a traditional thruster, propeller system. In addition, swimming techniques may improve the hovering and maneuvering capabilities of AUVs. Noiseless propulsion and a less prominent wake inspired by swimming creatures would have applications that would benefit military ocean vehicles [1].

Robotic fish such as the MIT Robotuna have explored the biology of aquatic creatures in order to better design underwater vehicles. An understanding of both the muscle structure and the swimming patterns of these creatures is needed to design such vehicles. This study focuses on the hydrodynamic mechanisms that result from the mode of swimming used by sea snakes and eels. This anguilliform motion varies from the motion of such fish as tuna and therefore this study provides insight into a different swimming motion.

Characteristics of Sea Snakes

Sea snakes are air breathing reptiles unlike eels with gills and are found in the Indian and Pacific oceanic regions. They are highly venomous; therefore, few studies have been done on them and especially on their swimming behaviors [2]. Sea snakes have evolved a form to survive the ocean.

The body is laterally compressed, starting out cylindrical and flattening towards the tail. They possess a strong flattened paddle-like tail which is believed to be the primary source of thrust. It is flat and wide up to its tip while the tail of a land snake is rounded and tapered toward the tip making the propulsive force generated by the sea snake larger than that of a land snake [2–5].

Fish Swimming

Fish swim with a variety of different motions. One way to classify fish swimming divides them into fish that move their body and or caudal fin (BCF) and those that use appendages, median and/or a paired fin (MCF). The BCF fish can be further subdivided into anguilliform, subcarangiform, carangiform, thunniform, and ostraciform swimmers [1]. These fish vary from highly undulatory (anguilliform) to highly oscillatory (ostraciform). In undulatory swimming, a wave passes along the propulsive structure. In ostraciform swimming, the propulsive structure instead pivots at a point. Undulatory BCF swimming uses a propulsive wave that travels in the direction opposite to the direction of movement of the fish. The wave travels at a speed greater than the swimming speed [1].

When fish swim, there is a transfer of momentum between the fish and the water through drag, lift, and acceleration reactive forces. A fish propelling itself at constant speed demonstrates momentum conservation. The thrust of the fish on the water equals the resistance of the water on the fish. This is referred to as self-propulsion. Drag can be subdivided into friction drag, form drag, and induced drag. Friction drag is caused by the skin friction between the fish and the boundary layer of the water. This is due to the viscosity of water in areas of large velocity gradients. Form drag results from pressures that arise when water is pushed out of the way during swimming. This is shape dependent. Streamlined bodies function to reduce form drag. Induced drag results from energy lost in vortices formed by the caudal and pectoral fins as they generate lift or thrust. The induced drag is dependent on the fin shape [1]. Through the study of fish shapes and swimming motions, ways to reduce these types of drag will emerge allowing lower drag underwater vehicles to be developed.

Anguilliform Swimming Motion

Anguilliform swimming has evolved separately in many vertebrate taxa and is found in ecologically and morphologically diverse fish, amphibians, and reptiles [6]. Anguilliform locomotion is also used by sea snakes, eels, lampreys, loaches, and gobies. Waves of shorter length than the body of the animal are propagated posteriorly along the length of the animal, causing the animal to move forward. The whole body participates in these large amplitude undulations. At least one complete wavelength is present along the body. The wave has an increasing amplitude as it passes from the head to the tail. This mode of swimming has been shown to be used over a range of velocities. Snakes are capable of backwards swimming by altering the direction of wave propagation. Most previous work has focused on other types of fish swimming and not on anguilliform swimming; thus, little is known about this mode of propulsion [1, 3].

Cp/Uo relates the overall fish swimming speed to the wave propagation speed and is an indication of swimming efficiency or the propulsive effectiveness of the propagating waves [1]. Studies on sea snakes found that they swim with a Cp/Uo ratio of 1.37 [7]. For sea snakes, the wave speed tends to increase along the length of the body and may be due to the muscle activation increasing along the body of the snake [6]. Undulatory fish move through the water and affect the water that is at right angles to the direction of motion. This causes the water to move laterally which can form a vortex wake behind the trailing edge. Therefore, part of the fish's energy is wasted on the creation of a vortex wake. The rest of the fish's energy output is available for propulsion, producing movement at a velocity Uo, against the viscous resistance of the water [8]. For efficient swimming, the amount of energy spent on the vortex production should be small.

Lateral compression towards the trailing edge is often present in undulatory swimmers, which improves propulsive efficiency as it decreases the mass of water next to the body that must be accelerated for propulsion. The forces needed for this water acceleration decrease and so hydro-mechanical efficiency increases. The inertial force and the drag force related to the body undulations can be used to generate propulsive forces. By increasing the height of the tail, the inertial force is increased by the added mass in proportion to the square of the height. In large elongate animals, the Reynolds number is large enough to change the inertial effects of the fluid into hydrodynamic forces and moments [3].

Unsteady Motion of Fish Swimming

Fish swimming is a demonstration of unsteady flow control by propagating a traveling wave along their backbone. Fish swimming has been modeled in biomimetic work such as the RoboTuna. In the RoboTuna, the undulatory motion is predominant in the aft half of the body, a characteristic of thunniform swimming. Fish also illustrate vorticity control. The undulatory motion of the fish combined with the flapping tail motion creates a jet wake allowing them to propel themselves. This was demonstrated by Triantafyllou and Triantafyllou who showed with the RoboTuna that a caudal fin could create a jet wake [9].

Unsteady motion of a body can be beneficial for efficient flow control. Vorticity in the wake of a self-propelled body demonstrates that a propulsive jet is needed to counter the drag of the body. Manipulation of the wake has been shown to increase propulsive efficiencies. High propulsive efficiency has been associated with a downstream wake of two to four vortices per period arranged in a reverse Kármán street [11]. Gray found that an eel produced a reverse Kármán street in the wake [3].

Tokumaru and Dimotakis investigated oscillation control by a rotating cylinder. These cylinders were forced to have rotary oscillations in steady flow. Dye visualization results of these studies showed that for a non-rotating cylinder, the dye dispersed and mixed over a large area. A typical Kármán vortex street was produced in the wake. For the rotating case, a wake of about the same size as the dye field in front of the cylinder was produced. This suggests that control over a cylinder wake can be achieved by controlling the rotating motion. The addition of circulation into the flow may allow control of the wake dynamics. In the case of the rotary cylinder, the forcing reduces pressure (wake) drag. Therefore, adding circulation into the flow may allow control of wake dynamics. The rotating cylinder showed a reduced wake signature. This illustrates how forced oscillations, an example of unsteady flow control, can result in significant drag reduction [10].

Flow About Two-Dimensional Waving Plates

The undulatory motion of fish swimming can be reduced to a traveling wave (Figure 1). The components of a wave include the amplitude (A), the wavelength (λ), the phase speed of the wave (Cp), and the velocity of the flow (Uo). Cp/Uo is the wave phase speed non-dimensionalized with the free speed and is an important parameter used in wave studies.

Several experiments have shown the unique effects of a waving boundary. In 1974, Taneda showed experimentally that when a traveling wave has a phase speed greater than the freestream speed, no separation off the crest of the wave occurred. Flow over waving boundaries was shown to separate if the traveling wave speed was slower than the freestream fluid velocity. This was seen for both the crest and the trough position of the wave. However, when the phase speed exceeded the freestream velocity, the flow remained attached and did not separate at the crest or trough regions [12]. When flow separates form drag (pressure drag) is increased, thus when the phase speed exceeds the freestream speed and no separation occurs, is ideal for decreasing drag on a swimming body.

Zhang completed numerical simulations of a waving boundary and showed a reduction in turbulence intensity when the phase speed exceeded the freestream speed illustrating another important benefit of a wavy swim-

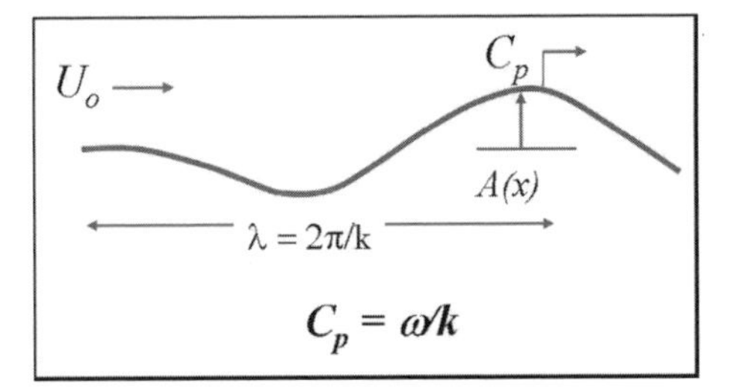

Figure 1. Drawing showing the components of a traveling wave. The freestream is Uo, the amplitude is A, and the phase speed is Cp.

ming body [13]. Zhang found that when Cp/Uo reached 1.2, there was significantly lower turbulence intensity. A decrease in turbulence intensity reduces frictional drag. Identifying a phase speed with reduced turbulence intensity would be ideal for designing low drag vehicles.

Techet studied near body turbulence intensity on a waving mat in the MIT Marine Hydrodynamics Water Tunnel. This work focused on a two-dimensional mat that spanned the entire tunnel test section. This experimental work showed that turbulence intensity was reduced for phase speeds up to 1.2 times the freestream velocity up to a Reynolds number of one million [14]. A minimum in turbulence intensity occurred for the trough and crest positions of the mat at Cp/Uo = 1.2.

The above described projects all focus on 2D bodies. In this work, experiments were performed to determine whether turbulence intensity is decreased for a 3D revolved swimming body and to examine its wake profiles.

DESIGN PROCESS

Examination of a Real Sea Snake

The initial stage in the design of the sea snake was the observation of a real sea snake of the species, *Astrotia stokesii,* at the Harvard University Museum of Comparative Zoology (Figure 2). These observations allowed for the visual understanding of key characteristics of sea snakes including lateral compression and a flattened tail. The flattened tail aids in propulsion by providing a larger surface area. These important characteristics were incorporated into the design of the model.

Three-Dimensional Model Design Process

A CAD model was made of the snake with Solidworks and then the individual parts that comprise the snake model were 3D printed with a

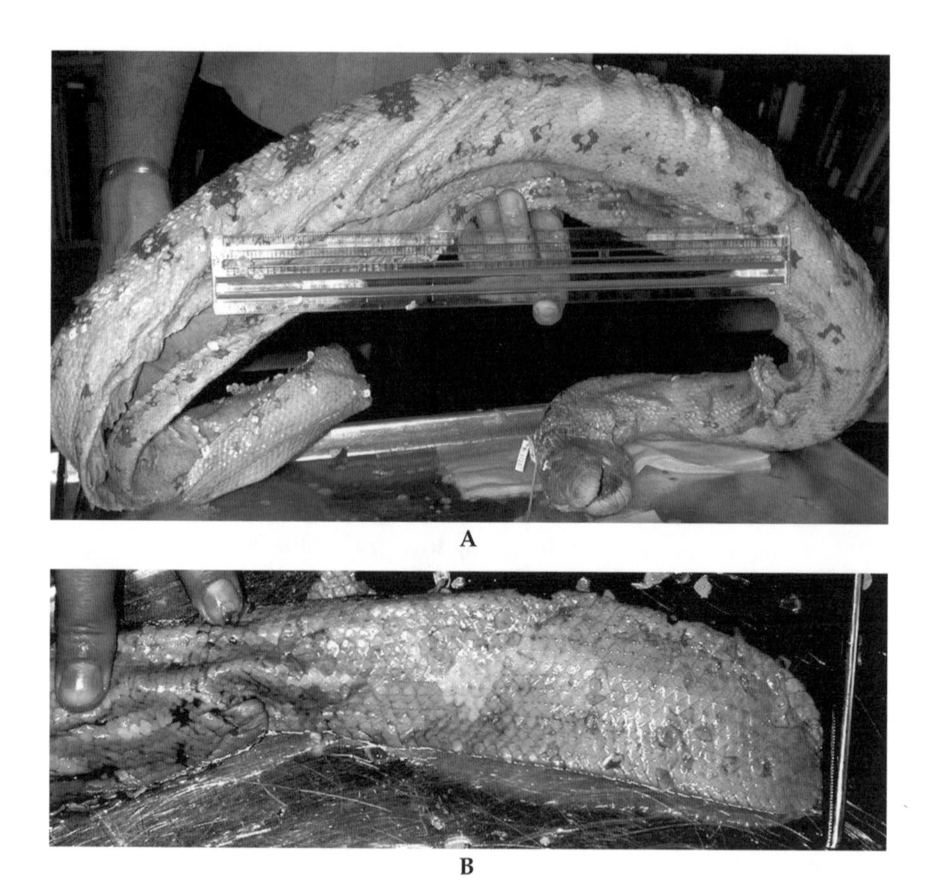

Figure 2. Photographs of the sea snake *Astrotia stokesii* taken at the Harvard University Museum of Comparative Zoology. **A** shows the entire snake and **B** shows the paddle-like tail.

starch/cellulose material and infiltrated with cyanoacrylate (superglue). The 3D printed parts fit together to form a single solid model one meter long. The model was composed of a head, a tail, and four body pieces. The head begins with a very rounded nose and then tapers to a constant elliptical diameter. The two snake parts aft of the head continue to display a constant elliptical cross-section. The fourth and fifth pieces gradually taper down towards the tail where piece five joins the flat paddle-like tail. The snake maintains a cross-section of 12.5 cm × 5 cm along the area with a constant cross-section. The tail then increases slightly in width but decreases in thickness. The tail is 23 cm long by 14 cm wide. The tail contains a central rib-like piece. Real snakes contain strong muscle in their tails to aid in the thrusting motion of the tail which is mimicked by this rib. In addition, since the tail is only 2.5 cm thick, the rib-like piece provides a point where a piston rod could be attached. This rib enables the snake tail to move in a constant thrusting motion.

Molding Process for Casting the Snake

The 3D printed model was cast with RTV Silicone to make a "negative" mold. The negative mold was filled with Evergreen 10 Smooth-on liquid urethane rubber to form the final snake. Evergreen 10 is a soft urethane material with durometer 10. It is low in viscosity which allows it to cure with negligible shrinkage [15]. At the locations where the piston rods attach to the snake, it was necessary to mold T-shaped connectors into the snake. These are later used to connect the snake to the crankshaft mechanism that drives the traveling wave down the snake. A metal plate was molded into the head to add both stiffness and to minimize the pitching of the head during the swimming motion.

EXPERIMENTAL SET-UP

MIT Marine Hydrodynamics Water Tunnel

The snake was installed in the MIT Marine Hydrodynamics Water Tunnel. The water tunnel is a 6,000 gallon recirculating water tank that spans two floors. An impeller drives the flow at speeds of up to 10 m/s. The test section is 1.5 m long and has a cross section of 0.5 m × 0.5 m. Each polycarbonate window of the test section is removable to allow for special test windows to be used. For the snake experimentation, the top window was replaced with the wave driving mechanism.

Figure 3 shows a photograph of the snake connected to the wave driving mechanism. The snake is located inside of the tunnel, connected to piston rods that drive the wave. The photograph identifies pistons five and six where the near-body turbulence studies were carried it. The two extreme positions of the snake motion, the trough and crest, are identified. For the data taken at a piston, when the snake is at its top-dead center location the snake at this point is in a "trough" configuration as viewed from the laser used in the LDV experiments which is located below the snake. When the piston is at bottom-dead center, the snake is considered to be in a "crest" configuration. The trough and crest positions are 180 degrees out of phase of each other.

The coordinate system shown in Figure 3 is used throughout the experimentation. The X direction is the direction of the incoming flow. The Z direction is represented by the up and down coordinate in the test section.

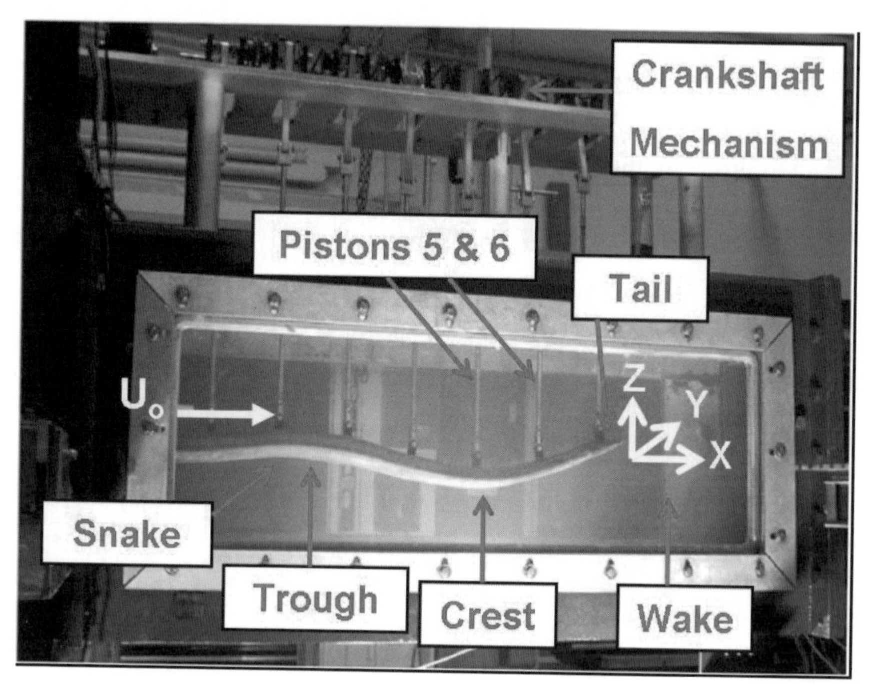

Figure 3. Photograph of the sea snake model installed in the MIT Water Tunnel. The inflow velocity is indicated. The snake is shown attached to the crank-shaft mechanism which drives the waves down the snake. Piston rods connect the snake and the crank-shaft mechanism. Pistons five and six are the locations where the near-body turbulence studies were completed. The tail and wake area are identified. The snake is shown in the two extrema positions, the trough and crest positions. The trough is when the snake is at its highest point and the crest is when the snake is at its lowest point.

The Y direction is the coordinate in and out of the test section. The zero point (0, 0, 0) is at the tip of the tail when the snake's tail is at its mean position.

Crank-Shaft Mechanism Used to Drive an External Wave

The snake was actuated by an external crank-shaft mechanism that was used successfully for Techet's 2D waving plate experiment [14]. Seven piston rods connect the snake to the crank shaft mechanism which drives an external wave along the snake. The pistons actuate the snake in the transverse direction and are located 13 cm apart from each other along the centerline of the snake body. The tail extends slightly aft of piston seven. Near sinusoidal motions of each piston location are created by cranks with varying phase angles which are mounted on a single drive shaft. The crankshaft mechanism was set up to allow the amplitude of the snake to increase linearly as the distance from the head of the snakes increased. The amplitude increased by the ratio x/16 where x is the distance in inches of the snake from the head. The nose of the snake was held fixed through a foil mounted to the top of the tunnel which attached to the internally molded metal plate.

Snake Motion Parameters

The snake undergoes a traveling sinusoidal wave shape with a linearly-increasing amplitude which is characteristic of a real snake. The snake's motion, y(t,x), is described by Equation 1, with the parameters shown in Table 1,

$$y(t,x) = A\frac{x}{L}\sin(kx - \omega t) \quad . \tag{1}$$

Table 1. Snake Motion Parameters.

y	lateral motion of the snake centerline
t	time
x	longitudinal coordinate of the snake, measured positive aft of the nose
A	amplitude of motion at the tail
L	length of the robot, 1m
ω	piston frequency (rad/s)
λ	wavelength of undulation
Cp	phase/wave speed

The phase speed, Cp, is defined as

$$Cp = \frac{\omega}{k} = f\lambda \ , \tag{2}$$

and the wavenumber, k, is

$$Cp = \frac{\omega}{k} = f\lambda \ . \tag{3}$$

The cranks were set to achieve a maximum amplitude of 6.25 cm at the tail and a motion wavelength, λ, equal to one body length or λ = 1.0 m. The amplitude of the snake increases linearly towards the tail with the amplitude equal to L/16. The tunnel water speed for all tests was 1.0 m/s; therefore, the Reynolds number based on length (Re_L) was approximately 1×10^6. The fixed crank arrangement leaves only one free parameter, the motor speed Ω which allows the value of Cp to be varied. The Cp values were varied from 0.4–2.4.

NEAR-BODY TURBULENCE STUDY AND WAKE VELOCITY CHARACTERIZATION USING LASER DOPPLER VELOCIMETRY

Overview of the Laser Doppler Velocimetry Technique

Laser Doppler Velocimetry (LDV) is a non-intrusive method that utilizes a laser and optic system to measure particle velocities. This technique measures mean velocities and turbulence by measuring a Doppler frequency shift. The fluid is seeded with small tracer particles that follow the flow. Light waves produced by the laser beams reflect when a particle passes through the intersection volume formed by the two coherent laser beams. A detector receives this scattered light that has components from both beams. The two intersecting beams form a fringe pattern of high and low intensity. The intensity is dependent on the difference between the optical path lengths of the two components. The fringes in space move across the detector at the Doppler difference frequency. When the particle traverses this fringe pattern, the scattered light fluctuates in intensity with a frequency equal to the velocity of the particle divided by the fringe spacing [16].

Experimental Set-up of the LDV System

A backscattering LDV system was used in combination with a traverse positioning system to make velocity measurements. A 6–Watt argon ion laser (continuous) was used in conjunction with TSI, Inc. optics. The laser is backscattering and reflects back off the particles to the receiving optics. The traverse system allows the laser to be positioned at 0.01 mm increments. A Dantec data acquisition system allows u and v velocities to be measured and then the data is averaged into angle bins based on the 360? of motion. Mean data was taken in the wake to determine the wake profiles.

In this experiment, the LDV measurements were synched with the body motion through an inductance encoder on the drive shaft. This allowed the data to be related directly to the swimming phase of the snake. The snake is run through many cycles allowing the data to be phase averaged over many cycles. The LDV gives the mean velocity components, $\bar{u}$ and $\bar{v}$, and allows for the calculation of the velocity fluctuations (u' and v'). Each near body turbulence data point represents ~36,000 measurements. For the wake, data from at least 30 cycles of the swimming motion was used.

Near Body Turbulence Measurements of the Snake

Near body turbulence levels were studied at two positions along the snake, at piston five and six (Figure 3). Piston five is a distance of 0.72L and piston six is a distance of 0.80L, where L is the length of the snake. Measurements were made at very small distances from the snake in order to assess turbulence production and control near the undulating body. Then data was taken at incremental steps away from the snake body all of the way into the freestream. Data was taken for both the crest and trough positions at pistons five and six for Cp/Uo values of 0.4, 0.8, 1.2, and 2.0. Data was bin-averaged into 360 bins of one degree each, with approximately 100 points per bin. Data points that were more than three standard deviations away were removed during data processing. Five of the bins were then combined to analyze the data.

The local phase locked velocity component (u) can be statistically split into two parts, the average velocity component u and its fluctuations u':

$$u = \overline{u} + u' \cdot \tag{4}$$

The standard deviation or rms of the velocity fluctuations can be used as a measurement of turbulence intensity. The standard deviation is calculated by

$$u' = \sqrt{\sum \frac{(u_i - \overline{u})^2}{n}} , \tag{5}$$

where n is the number of samples and u_i is the instantaneous velocity sample. The base level of turbulence is approximately 5%, in the absence of the snake apparatus, for speeds below 1 m/s.

In this section, turbulence intensity is shown for data taken at pistons five and six as a function of Cp/Uo for various distances. The axes on these graphs show turbulence intensity, $\overline{u}'^2$, normalized by inflow velocity vs. Cp/Uo. The turbulence intensity is represented in these plots by the symbol <uu>. The distances shown are the distances from the snake body.

Near Body Turbulence Results at Piston Five

The trough data (Figure 4) shows the highest intensity at positions closest to the snake body which is expected. Data taken too close to the snake body can result in the laser beam being clipped by the body; therefore, data closer than 4 mm to the body is not shown. This means that the laser is focused on the body and not on particles in the water and therefore, data is not recorded. As distance from the snake increases, the turbulence levels decrease. When 8 mm away from the snake is reached, a minimum for Cp/Uo = 1.2 develops. The low that develops for Cp/Uo = 1.2 remains visible until a distance of 18 mm. After this distance, the fluid turbulence level appears to stabilize. This fluid no longer feels the effect, or feels a minimum effect of the snake body. The fluid appears to reach the turbulence

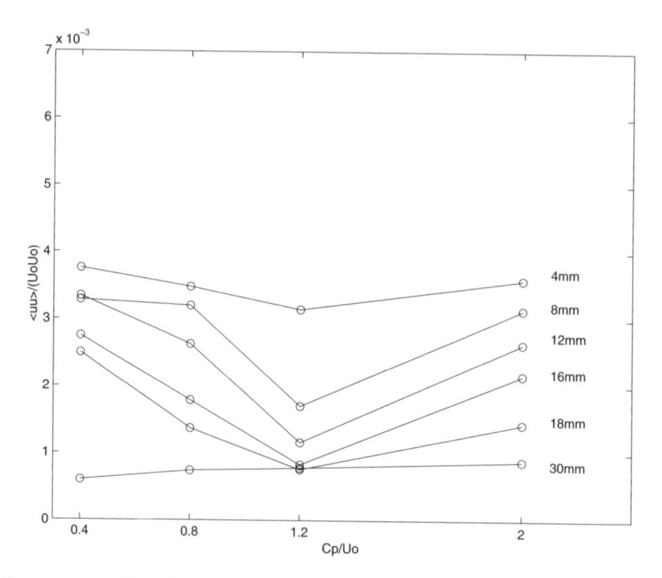

Figure 4. Summary of turbulence intensity at the piston five trough position. The turbulence reduction at Cp/Uo = 1.2 can be seen. This agrees with previous two-dimensional waving plate studies. At a distance of 30 mm away from the body, the turbulence level has reached the level of the freestream in the tunnel.

level of the freestream at a distance of 30 mm. The minimum at Cp/Uo = 1.2 is consistent with the results obtained by Techet.

The data taken when the snake was in the crest position is shown in Figure 5. The distances closest to the snake again show the highest turbulence. However, no minimum for Cp/Uo = 1.2 occurs. High points exist at Cp/Uo = 0.4 and 1.2 and low points at Cp/Uo = 0.8 and 2.0. The shape of this plot is unlike the results shown by Techet. The three-dimensional shape of the snake allows flow to go over the sides of the snake. This does not occur for Techet's mat as it spanned the entire width of the tunnel. The flow over the sides of the snake may affect the near body turbulence levels. Also, since the snake does not span the tunnel, there is an increased effect of the pistons in creating turbulence.

Near Body Turbulence Results at Piston Six

Turbulence intensity data taken at the crest and trough positions for piston six is shown in Figures 6 and 7. For both the trough and crest positions, the highest turbulence is located close to the snake body. Within the trough, as distance from the snake increases, turbulence levels decrease for all Cp/Uo. As the distance from the snake further increases, a minimum for Cp/Uo = 1.2 appears. Turbulence

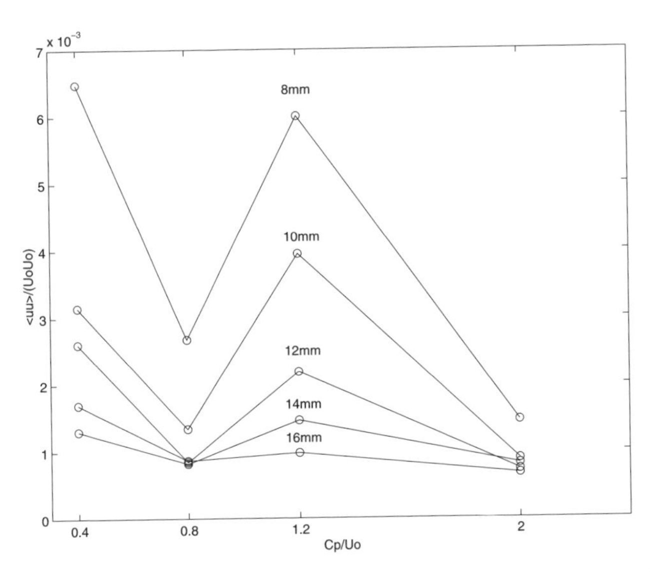

Figure 5. Summary of turbulence intensity at the piston five crest position. The turbulence maxima at Cp/Uo = 1.2 can be seen. This is unlike the two-dimensional case which shows a turbulence minimum. At a distance of 16 mm away from the body, the turbulence level has reached the level of the freestream in the tunnel.

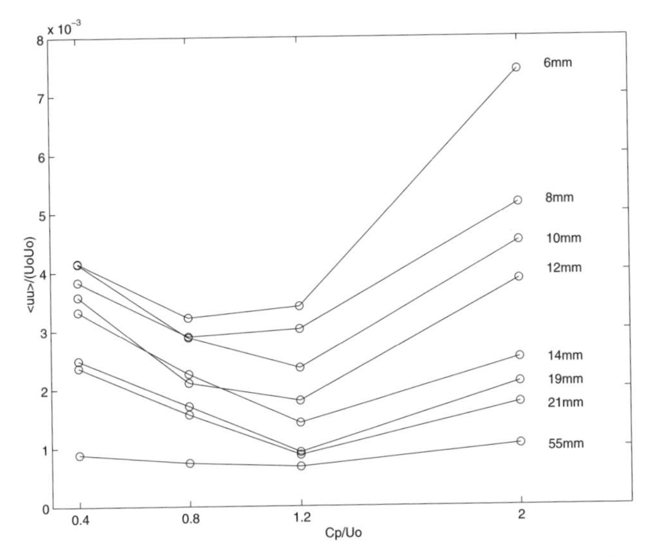

Figure 6. Summary of turbulence intensity at the piston six trough position. A turbulence minimum appears for Cp/Uo = 1.2 starting at 12 mm away from the body. The minimum for Cp/Uo= 1.2 agrees with the two-dimensional studies. At a distance of 55 mm away from the body, the turbulence level reaches the level of the freestream

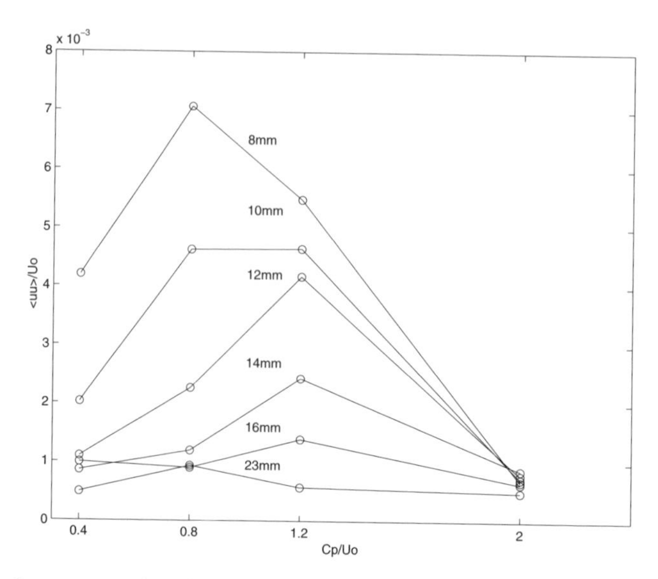

Figure 7. Summary of turbulence intensity at the piston six crest position. A turbulence maximum appears for Cp/Uo = 1.2. This is an opposite effect than is seen for the crest position in two-dimensional studies. At a distance of 23 mm away from the body, the turbulence level has reached the level of the freestream.

intensity reaches tunnel levels of three percent for all Cp/Uo at a distance of 55 mm from the snake body. At the crest position, a decrease in turbulence is again seen as the distance from the snake increases. However, now a maximum is seen in turbulence intensity for Cp/Uo = 1.2 for 12 mm–16 mm. This is opposite the case for the trough where Cp/Uo = 1.2 shows the lowest turbulence in the crest. Tunnel turbulence levels are reached at 23 mm.

Turbulence reduction is demonstrated for this 3D case for the trough position which supports previous data for 2D waving plates. However, in the 2D case, the crest position shows a minimum at Cp/Uo = 1.2 unlike the 3D case where a maximum occurs. Again, the differences could be a result of three-dimensional effects of flow around the snake body which is unlike Techet's flat plate that spanned the entire section of the tunnel, preventing flow around the edges. In addition, this snake data was taken further downstream at piston six where the amplitude is larger.

CONCLUSIONS

This work examines the hydrodynamics of a swimming sea snake model. This project demonstrates the ability to study the hydrodynamics, both near-body and wake, of anguilliform swimming motion with the use of the snake model developed and the MIT Water Tunnel. Two types of measurements were made, velocity and pressure. The experiments demonstrated the effectiveness of kiel probe and LDV technology for studying hydrodynamics in the water tunnel.

Using biomimetic principles, a model sea snake was designed. The snake was cast with polyurethane rubber and was constructed to use Techet's waving plate mechanism. Tests in the water tunnel showed that the snake model was capable of being used for characterizing the hydrodynamics of a three-dimensional swimming body.

Near-body turbulence was studied with an LDV system. Measurements were taken at pistons five and six with the snake in both the trough and crest position. At both piston locations, turbulence reduction was shown for the trough position for Cp/Uo = 1.2. This supports previous two-dimensional studies, both experimental and numerical. The crest position, however, showed the highest turbulence at Cp/Uo = 1.2. This differs from the two-dimensional studies which exhibited a turbulence minimum in the crest at this Cp/Uo value. This near-body turbulence data raises important questions about the flow of the 3D snake vs. a 2D plate.

ACKNOWLEDGEMENTS

I would like to thank the Link Foundation for their generous support of my studies through their fellowship program. I would like to thank my advisor, Professor Michael Triantafyllou, for all of his time, answers, and support. I would like to thank Dr. Franz Hover and Professor Alexandra Techet, a past Link Foundation Fellow, for all of their help during the many stages of this work. In addition, I would like to thank Dr. Richard Kimball, the many UROP students involved with this project, and the students of the MIT Towing Tank, especially Joshua Davis and David Beal, for all of their assistance. There are several people that were key to the making of the snake: Jose Rosado, Curatorial Associate, at the Harvard University Museum of Comparative Zoology showed me real sea snakes, Mike Tarkanian 3D printed the snake at Z-Corp, and Fred Coté at the Edgerton Student Machine Shop helped me to machine many parts. This work was funded by DARPA.

REFERENCES

[1] M. Sfakiotakis, D.M. Lane, and J.B.C. Davies, "Review of fish swimming modes for aquatic locomotion", *IEEE Journal of Oceanic Engineering* 24(2), 237–252, (1999).

[2] H. Heatwole, *Sea Snakes,* Krieger Publishing Company: Malabar, FL, (1999).
[3] A. Azuma, *The Biokinetics of Flying and Swimming,* Springer-Verlag: New York, (1992).
[4] W.A. Dunson, *The Biology of Sea Snakes,* University Park Press: Baltimore, MD, (1975).
[5] J.B. Graham and W.R. Lowell, Surface and subsurface swimming of the sea snake *Pelamis platurus, Journal of Experimental Biology,* 127, 27–22, (1987).
[6] G. Gillis, Undulatory locomotion in elongate aquatic vertebrates: anguilliform swimming since Sir James Gray,*American Zoologist,* 36, 656–665, (1996).
[7] J. Videler, *Fish Swimming,* Chapman and Hall: London, (1992).
[8] J. Lighthill, Hydromechanics of Aquatic Animal Propulsion—A survey, *Annual Review of Fluid Mechanics,* 1, 413–446, (1969).
[9] M. Triantafyllou and G. Triantafyllou, An efficient swimming machine, *Scientific American,* 272(3), 64–70, (1995).
[10] P. Tokumaru and P.Dimotakis, Rotary oscillation control of a cylinder wake, *Journal of Fluid Mechanics,* 224, 77–90, (1991).
[11] M. Triantafyllou, G. Triantafyllou, and D. Yue, Hydrodynamics of fish-like swimming, *Annual Review of Fluid Mechanics,* 32, (2000).
[12] S.Taneda and Y.Tomonari, An experiment on the flow around a waving plate, *Journal of the Physical Society of Japan,* 36(6), 1683–1689, (1974).
[13] X. Zhang, I. Surfactant effects on the interaction of a three-dimensional vortex pair with the free surface; II.Turbulent flow over a flexible body undergoing fish-like swimming motion, PhD Dissertation, Massachusetts Institute of Technology, Department of Ocean Engineering, (2000).
[14] A.H. Techet, Experimental visualization of the near-boundary hydrodynamics about fish-like swimming bodies, PhD Dissertation, Massachusetts Institute of Technology and the Woods Hole Oceanographic Institution, Department of Ocean Engineering, (2001).
[15] Smooth-On Technical Literature.
[16] M. Sommerfeld and C. Tropea, Single-point Laser Measurement, *Instrumentation for Fluid-Particle Flow,* 252–317, (1999).